what to
expect
when you're
expecting
robots

THE FUTURE OF HUMAN-ROBOT COLLABORATION

what to expect
when you're
expecting
robots

LAURA MAJOR
and
JULIE SHAH

BASIC BOOKS
New York

Basic Books
Hachette Book Group
1290 Avenue of the Americas, New York, NY 10104
www.basicbooks.com

Printed in the United States of America
First Edition: October 2020

Published by Basic Books, an imprint of Perseus Books, LLC, a subsidiary of Hachette Book Group, Inc. The Basic Books name and logo is a trademark of the Hachette Book Group.

The publisher is not responsible for websites (or their content) that are not owned by the publisher.

Print book interior design by Linda Mark.

Library of Congress Cataloging-in-Publication Data
Names: Major, Laura, author. | Shah, Julie, author.
Title: What to expect when you're expecting robots: the future of
  human-robot collaboration / Laura Major and Julie Shah.
Description: First edition. | New York: Basic Books, 2020. | Series:
  The future of human-robot collaboration | Includes bibliographical
  references and index.
Identifiers: LCCN 2020026201 | ISBN 9781541699113 (hardcover) |
  ISBN 9781541699106 (ebook)
Subjects: LCSH: Robotics—Human factors. | Robots—Forecasting. |
  Human-machine systems—Social aspects. | Technological forecasting.
Classification: LCC TJ211.49 .M35 2020 | DDC 629.8/924019—dc23
LC record available at https://lccn.loc.gov/2020026201

ISBNs: 978-1-5416-9911-3 (hardcover); 978-1-5416-9910-6 (ebook)

LSC-C

10 9 8 7 6 5 4 3 2 1

*Dedicated to our children, Luca, Lily, Lillian, and Vivien,
who inspire us to make a better world.*

Contents

Introduction

IMAGINE A WORLD FULL OF ROBOTS. IT'S A BIT LIKE THE world of today, except these robots are not just expensive novelties. They aren't limited to a small handful of jobs and don't need you to tell them what to do. Instead, these robots are something like partners— they cooperate with you the same way teammates on the basketball court cooperate with each other. One sets a pick so the other can roll; one lobs the ball high above the basket and another swoops in to dunk. We call this *human-robot collaboration*, and it is likely to revolutionize our relationship to technology over the next several decades.

People seem to be concerned about whether robots will one day make us obsolete—whether they will become smarter, faster, better than their human creators. But the reality is that robots and humans will probably always be good at different things. And, as we intend to show here, it is possible that some of our most stubborn societal problems could be better addressed by the kind of collaboration we envision. The applications are vast. Through a symbiosis of human and artificial intelligence on the road, we can dramatically reduce fatalities due to car accidents and start to tackle the congestion problems that plague nearly every city in the world. Robots as personal augmentation systems can improve daily well-being and enable us to thrive independently far into old age and as our abilities change. Robotic orderlies can make emergency rooms safer and more efficient, shortening wait times and enhancing care. Robots will bring countless other small but meaningful improvements to our daily lives, and as working moms with an ever-growing to-do list, we personally look forward to these changes.

You may be thinking that robots like this are already here. After all, your Roomba can vacuum your living room on its own. But while your Roomba's ability to map your floor plan might seem impressive, it is not really much different from any other household appliance. And this is true of most robots we currently encounter. We restrict their roles with simple rule-based behaviors and interact with them through taps on screens and other simple commands. They understand very little about us, and we request relatively little of them in return. Robots in factories work in cages. We turn on adaptive cruise control for our commute, but turn it off as soon as we hit traffic. We wake up in the morning and ask Siri about the weather, or tell Alexa to add milk to the shopping list, but ultimately, we dress ourselves and buy our own milk. We don't judge our Roomba too harshly when it gets stuck on a tuft of carpet or misses spots. It is a simple machine, not a very smart one. Most robots today have narrow functionality, can only operate in controlled environments, and require essentially constant human oversight. And given those three limitations, they perform beautifully.

But human-robot collaboration is something altogether more revolutionary. New types of intelligent robots are just now beginning to enter our cities and workplaces, and they are defined in large part by the way they transcend these limitations. Robots making package deliveries in our neighborhoods, or shopping for us at grocery stores—what we call *working robots*—can no longer be considered mere *tools*. They amount to new social entities. Let us be clear: whether these robots can be said to be conscious or as intelligent as humans is not really the point, and indeed, many working robots will be a far cry from sentient. What we need to understand about tomorrow's robots is that they are going to be something *different*, with roles mediated at all stages by the rules of social interaction. They will become more human in one specific way: whether they make our lives better or worse comes down to whether they know how to behave.

And there are many of them coming. If the picture we are painting feels like a far-off dream, it is because we are sleeping. There are 1.7 million industrial robots in operation around the world today.[1] That's the same number as the human population of Boston, Pittsburgh, and San Francisco combined. There are currently 30 million robots in our homes in the United States.[2] And that's not counting the Alexas, Siris, smart home devices, sidewalk delivery robots, grocery store robots, apartment security guard robots, and hospital service robots now making appearances as we visit friends, run errands, and go shopping. Soon, your front yard and our neighborhoods may be swarming with drones. The National Aeronautics and Space Administration (NASA), entrepreneurs, and industry leaders are working quickly to open up our skies to urban air mobility—where drones deliver small packages and passengers zip across town in the air over roadway traffic.

IT IS TUESDAY MORNING. YOU STEP OUT OF YOUR HOUSE AND WALK toward your car, which is parked on the street. Meanwhile, a delivery robot is zipping down the sidewalk, attempting to deliver packages in time for the single-day shipping deadline. It detects you as a nearby obstacle and stops for safety, but not in time. You snag your foot on it and lurch forward, catching your fall. It's already a bad way to start a day. As you drive to work, you pause for a pedestrian to cross the street, and just as you proceed you spot a small assistant robot, probably carrying the person's laptop and lunch, trailing behind. It's low to the ground, like a puppy—and you almost didn't see it. You slam on the brakes, narrowly avoiding the robot with its cargo of laptop and sandwich, but the car behind you gently rear-ends you. After you thankfully confirm there was little damage to either of your bumpers, you get back on the road, but you feel the frustration grow when you realize you are now behind an autonomous vehicle doing a test run through a new neighborhood. It feels like time has slowed while you pace behind the vehicle at just below the speed limit, stopping for every object within ten feet regardless of whether the object is actually on the road. You're ready to pull your hair out by the time you arrive at the office, only to be stuck at the elevator bank by a delivery robot that you can't get to move out of the way of the buttons. It is only 9:00 a.m., and four different robots have already made your life more difficult, simply because they have not been designed to understand or care about you. You can't help but wonder who these robots are really helping.

How can you make a robot that understands strangers? The way to make this work is not to build arbitrarily "smarter" or "more powerful" robots, but to rethink what it is we expect from technology. Consider the search-and-rescue dog, for example. The dog doesn't have to be commanded so much as it is guided by its human handler—with subtle hand gestures indicating areas for focus of attention. The dogs often act on their own. Their handlers depend on them, and the dogs have their own set of social norms for interacting with people. The vests they wear remind people not to touch or interact with them. In challenging spaces, they are put on a leash, so that their behavior might be more tightly controlled. They are still dogs, but because their roles—and their handlers' roles—have been carefully designed, search-and-rescue teams are able to do much more than either a dog or a person can do independently.

We envision human-robot collaboration in this way: people and robots buzzing around each other, sometimes working as individuals, and other times collaborating in groups. With a wave of a hand we will be able to offload a burdensome task to a robot, and people in need might call on multiple robot helpers to accomplish things they would not be able to accomplish on their own. But getting this right is a two-way street. Just as robots will need to be able to understand social norms, they will also require us to reconsider the place of technology in our everyday lives. We will have to make certain changes, individually and as a society, to incorporate robots into our world. This partnership will require new human and robot languages and norms. We will have to rethink our infrastructure. We will have to consider the implications of the fact that these robots will be commodities, available to some and not to others. And because all this is being set in motion by a handful of tech companies, we will need to be clear about industry's ethical responsibilities. It will take deliberate, collective action. That is the purpose of this book: to figure out what makes a robot socially and personally valuable, and then consider how we as a society can ensure that the robots we make have these characteristics. Robots will force us to rethink the role of technology in society—from the practical level, such as figuring out how robots will deal with bystanders as they make their way through our neighborhoods, to more philosophical issues, especially navigating the tension of how these technologies will differentially impact groups in society. The tech industry is currently leading this change, but the arrival of autonomous robots in society impacts all of us. In order to fully embrace this future, it will require whole-society efforts.

Over the past few years, Julie's husband—a physician—has frequently texted her pictures of new robots he has encountered in the hospital where he works. One day a new one appeared, delivering medications floor to floor. A medical doctor entered an elevator to find one of these robots with a sign on it: "DO NOT GET IN ELEVATOR WITH ROBOT." The hospital staff was not supposed to come in close contact with the robot or interact with it in any way because it was still learning how to function safely and effectively in the human environment. Occasionally, the robot could lose its sense of where it was within the hospital, and would stop or start up again abruptly as it searched for clues. Naturally, this unpredictable behavior would be a concern in a confined space like an elevator. The robot was still basically a student, and just like a student driver, it needed to be treated with caution.

We know instinctively how to change our driving behavior when we see the "Student Driver" sign on a car. Good driving is not just about knowing the rules contained in a handbook; it involves the many informal rules and behaviors that drivers can only really learn with experience. That drivers know and abide by these rules make driving (for the most part) predictable. Student drivers do not yet have these mental models. Their ignorance, and perhaps their nerves, too, make them somewhat unpredictable, which may create an unsafe environment. The sign means that every experienced driver around that car should expect the unexpected.

Imagine how exhausting and stressful it would be if you had to drive every day surrounded by student drivers. Now imagine how exhausting and stressful—and possibly dangerous—it's going to be to coexist with hundreds of robots in our lives every day, whether on roads, in the hallways of our office buildings, in our parking lots, in our restaurants or hospitals, or buzzing around overhead as we walk down the street, especially if they don't understand the rules we all follow that make those spaces navigable and safe. The fact of the matter is that, at some point soon, robots will not yet be like "people," but neither will they be strictly "tools," only moving when we command them. They will be something altogether new. But what will that be? We believe that tomorrow's societies will be run increasingly by a new kind of relationship with technology, a *human-machine partnership*. The upside is tremendous. Consider that car accidents cause nearly 1.25 million deaths annually across the globe.[3] That's over 3,000 fatalities on our roadways every day, about 100 per day in the United States alone.[4] It is the ninth leading cause of

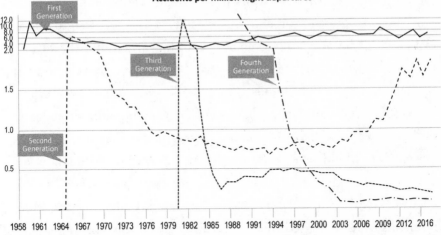

FIGURE 1: Accident rates for four levels of technological innovation in airplane technology: first generation (early commercial jets), second generation (more integrated auto-flight), third generation (glass cockpit and flight management system, or FMS), and fourth generation (fly-by-wire). *Source: A Statistical Analysis of Commercial Aviation Accidents, 1958–2016* (Blagnac Cedex, France: Airbus, 2017), https://flightsafety.org/wp-content/uploads/2017/07/Airbus-Commercial-Aviation-Accidents-1958-2016-14Jun17-1.pdf.

death worldwide. And yet there were only 500 deaths due to commercial aviation accidents in 2018 and only 144 fatalities the year before.[5] This is in part because the aviation industry has already embraced the idea of a human-machine partnership. Over the past few decades, the industry has reconceptualized the relationship between the pilot and the aircraft, and as a result, each is able to compensate for the difficulties of the other, and the skies are safer. Automation offers us the opportunity to reach similar, seemingly unachievable safety results on our roadways. Imagine a world in which fewer than 100 people die annually from traffic accidents around the world. With human-robot collaboration, such a future is possible.

Achieving the safety standards that have made aviation so safe and reliable hasn't been easy. But the hard lessons learned offer some guidance as we begin to think about introducing robots into our lives that we can trust not to cause real harm. Because, make no mistake: although some independent robots will be doing work that is potentially only annoying or inconvenient, others will have the potential to hurt or even kill people. Any autonomous system unleashed into the chaotic realms of our modern lives without careful consideration of how it will impact us could be dangerous. In this book, we will look, by turn, at the kinds of decisions we will have to make when building a new social entity. Probably the most important lesson we can offer is this: people design robots, and people are imperfect. Each time automation systems were introduced into cockpits—autopilot, glass cockpit, and fly by wire—fatal accidents temporarily increased. Only after the initial problems were addressed were we able to reap the benefits. No matter how hard an engineering team tries to properly design a new system or how rigorously evaluators and regulators test it, it can never be perfect.

Through decades of hard work, experimentation, and refinement, engineers have optimized the complex human-technology partnership that makes our commercial air transportation system work for our benefit and well-being. But, much like parenthood, a human-technology partnership takes work. It is not something we can ever expect to be perfect right out of the gate. Think about the learning curve for pilots today. They still have to train extensively on the logic and behavior of the automation, and learn how to communicate with and rely on the automation. But they also train to maintain the skills to manually fly an aircraft in case the automation fails. The partnership between a pilot and the flight management system is honed through this training. The automation can't be preprogrammed to work perfectly with the pilot, in much the same way a person can't be programmed to live in a happy marriage with another human being. Developing automated systems takes time, expertise, and financial investment. But how many of these working robots can we reasonably invest in? There will be many robots performing many more tasks than we can imagine today; they will be navigating our everyday world as best they can; and we will have to work with them as best we can without always having the luxury of extensive training or know-how. Still, the lessons from aviation offer insight into how robots and humans can work together, and we'll discuss many of these lessons at length. The main point remains: human-robot collaboration will force us to conceptualize new ways of integrating technology into society.

Determining what makes an effective human-machine partnership is trickier than it initially sounds, especially when safety matters. Imagine the dangers of negotiating throngs of robots zipping along our sidewalks as we walk down city streets. The problem is not just a matter of scale, of the increasing number of robots in our everyday lives, or of the physical proximity to more robots. Instead, the core of the challenge is a shift in the nature of consumer automation itself—from accessory technology to safety-critical systems.

Figure 2 illustrates what we mean. Here, we have attempted to capture the cost of failure across different applications and the amount of training that is required for their operation. Industrial applications, such as commercial flight, are highly complex: it is not possible to safely control these systems without robotics. Furthermore, the operators of these systems are highly trained, not only on the fundamentals of the applications (such as physics, aerodynamics, and electromechanical systems), but also on the robot. This training gives them the knowledge they need to manage the system even in the face of failures. Industrial applications are represented with squares in the figure.

Consumer products, in contrast, do not historically pose any serious safety risks and are typically designed to be used out of the box without training (beyond, perhaps, reading the instruction manual). Examples include Siri, Roombas, and Alexa. These are represented in the figure with diamonds.

Collaborative working robots represent a new class of consumer products that fall somewhere in between traditional consumer products and industrial applications. They introduce robotics into safety-critical

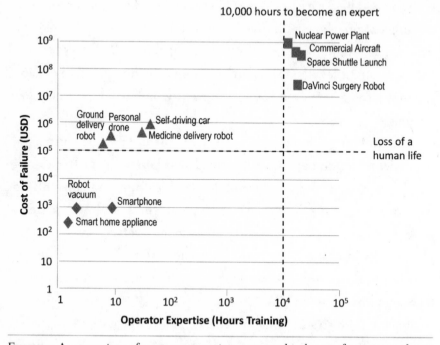

FIGURE 2: A comparison of operator expertise, measured in hours of training and cost of failures, for three classes of applications: industrial applications (squares), commercial products (diamonds), and a new class of safety-critical commercial products (triangles). *Source*: Laura Major and Caroline Harriott, "Autonomous Agents in the Wild: Human Interaction Challenges," in *Robotics Research: The 18th International Symposium ISRR*, ed. Nancy M. Amato, Greg Hager, Shawna Thomas, and Miguel Torres-Torriti, Springer Proceedings in Advanced Robotics, vol. 10 (Cham, Switzerland: Springer, 2020).

activities and areas—self-driving cars on our streets, delivery drones on our sidewalks, robots monitoring inventory or cleaning up spills in grocery stores, and medication-delivery assistants in our hospitals. Whether such robots can survive in the real world is ultimately a question of whether we can figure out how to interact with them. We already interact with consumer technologies, but how we interact now will not sufficiently account for all the new, potentially dangerous things these robots will be capable of doing in the future. Surely, they cannot be designed using exactly the same process as cockpit or spacecraft automation—we cannot afford to have all of us become experts about every robot we encounter, as pilots do with flight training, and anyway, everyday life is a lot harder to predict than a flight pattern. Getting the new hybrid design process right will be the key to unlocking an exciting and productive future where humans and machines truly partner to enhance our lives, and where we will each draw strengths from the other's capabilities. Getting it wrong is, for all the damage it will do, simply not an option.

The second major challenge is that this new class of safety-critical consumer products cannot be designed without careful consideration of social norms, just as aircraft cannot reasonably be designed without consideration of the air transportation system infrastructure and constraints. Aircraft are designed with special equipment to communicate with air traffic control and other aircraft. Pilots must have their flight paths approved before takeoff, and they must request a change and receive approval before modifying that path during flight. Their navigation solution is based on acceptable options defined by regulatory bodies, and different rules apply to aircraft with different capabilities. Some must stay at lower altitudes and only fly on clear days; others are allowed to fly when visibility is extremely low and can get closer to other aircraft as they pack into smoother, faster tracks, reducing flight time and delays. Similarly, working robots will need to work within social norms and abide by rules and regulations that guide their safe use across societal situations. We must embrace this change and be thoughtful about how to design robots as partners, creating the necessary infrastructure and support for these new entities rather than having to relearn the hard lessons we encountered in the early decades of air transportation. In other words, we must begin to understand robots as *sociotechnical* systems.

The central goal of this book is to explore how the unprecedented design challenges of tomorrow's working robots force us to confront a central question about the role of technology in society: What can we expect from machines, and what can they expect from us? Out of this central question come several ideas that will structure our tour:

How can we harness the relative strengths of humans and robots? Can autonomous systems be too independent for their own good? How do you plan for the bystanders who will be subjected to other people's systems working in public spaces? We provide design frameworks and solutions for how humans and machines can better predict each other's behavior. As a key to our approach, we introduce the notion of *automation affordances*, that is, design features that provide clues to the user or bystander as to how they might be able to influence or adjust a robot's behavior. More than natural language capabilities, automation affordances will provide for a basic language for robots and people to communicate with each other in any encounter. The introduction of working robots, as a social concern, will require a level of transparency and collaboration that is currently mostly absent from the tech industry. We will discuss new methods of evaluating and testing intelligent machines in order to ensure that profit motives do not put public safety at risk. We will also discuss the limitations of conceiving of these challenges as solely technological problems, and the ways in which we will need to co-design our society for working robots.

While there is much we can learn from this broader perspective— taken from aerospace, industrial systems, and human-systems engineering—we also address the ways in which we are now grappling with challenges of human-robot collaboration that have no precedent in other industries. Working robots are safety-critical consumer products operating within environments that are less controllable and predictable than the applications in these other arenas, and they will be interacting with people who can be expected to have little or no training with them before they arrive on the scene in large numbers.

In this new setting, our old paradigms for using and working with technology quickly fall away. It feels quite comfortable to us to command our simple robots today, because we are clearly in control. But the world is changing rapidly around us. Robots are quickly evolving:

instead of mere tools that we command and query, they are becoming intelligent partners that we will work with. We know this because it's our job in academia and industry to drive the advancement in artificial intelligence and robotics to make this vision a reality. Julie leads an artificial intelligence (AI) and robotics research lab at the Massachusetts Institute of Technology that focuses on the future of work and the potential for reverse-engineering the human mind to make robots better teammates. She has pioneered new forms of human-robot teaming in manufacturing, transportation, and health care. Laura has been leading teams in industry to design and develop new autonomous systems for our skies and roadways, revolutionizing digital assistants on our battlefields and bringing autonomous cars into reality. We are aided in this undertaking by unprecedented levels of private and public investment in new intelligent robot technologies, with visions of smart cities, schools, and workplaces on the horizon.

Along with these visions of possible futures, many of us feel great unease about the potential economic, workforce, and societal implications of these new technologies. We see billboards on our roads by insurance companies urging us to prepare for retirement because the robots are coming. Many of us worry about automation taking over our jobs, or about the *singularity*: the moment in the future when we enter an era of exponential growth in technology due to advances in AI.

It will surely require a collective effort to ensure that the new intelligent robots are harnessed to the task of enhancing human well-being. But here is the central contention of this book: robots need not be superintelligent, bent on world domination, or capable of making the entire human workforce obsolete to pose a threat to human prosperity. If they don't know how to behave in public, that will be enough. A robot that knows how to ride the subway could be truly revolutionary. One that doesn't, and tries to catch a train anyway, could do more than just make a few people late for work.

There is a big difference between functional use of technology and acceptance of robots en masse: one sidewalk delivery robot may be a novelty, whereas hundreds of them blanketing your city could be a dangerous prospect. We are sitting at the precipice of this phase transition, and we must begin to shift our conversation from one of fear to one of solutions. Those solutions live at the intersection of society and technology. It takes a village to raise a child to be a well-adjusted member of society, capable of realizing his or her full potential. So, too, a robot.

CHAPTER ONE

The Automation Invasion

F OR AS LONG AS ROBOTS HAVE BEEN IMAGINED, WE'VE WON-
dered not only what they can do, but what they should do. Yet such
debates have always seemed a bit academic, because in our daily lives
they have started popping up around us without much notice. It feels
as if we are always playing catch-up, never with enough lead time to
think deliberately about the technology and its role in our lives. Autonomous cars seemed to just appear on our roads one day. We crossed the
line to a tipping point in technology where automated driving suddenly
seemed within reach—so industry, government, and venture investors
went for it, taking the rest of us with them. Robots have started to
appear in grocery stores, gliding up and down the aisles looking for
spills and other safety hazards. The customers in the Stop and Shop
we visit in suburban Boston take it all in stride, for now at least. But
supermarket employees wonder how quickly the roles of these robots
will expand.

The emergence of new technologies can often feel abrupt, almost magical. This is because most of us are unaware of the many years of often incremental, commercially uninteresting technological innovations that make those breakthroughs possible. Occasionally a headline will herald some new conceptual development, which will quickly recede into the background of most people's lives. It may take decades before a robot appears that turns those technical innovations into a marketable machine.

Self-driving cars are our bellwether for the emergence of a new class of intelligent robots to be unleashed on society. They mark one of our first opportunities for everyday people to grapple with questions of sharing decision-making authority and control with an autonomous system. But Mercedes-Benz developed one of the first self-driving cars as early as the 1980s, and it could navigate streets without traffic at speeds of up to thirty-nine miles per hour.[1] Then, through the 1990s and early 2000s the US government funded the development of increasingly capable autonomous ground vehicles for military applications, culminating in the Grand Challenges sponsored by the Defense Advanced Research Projects Agency (DARPA).[2] In these groundbreaking events, cars drove autonomously for more than one hundred miles, including in urban environments. From about 2005 on, the automotive industry seriously invested in bringing self-driving cars into reality. Tesla introduced the first autopilot technology commercially in 2015.[3] Of all those stories, only the Tesla and its class of competitors seemed to sustain the public's attention.

A thirty-year time frame from initial headline to commercially viable product seems like a reasonable heads-up. But the story of how today's autonomous car came to be starts even further back in history, in aviation and industry. An aviation pioneer, Lawrence Sperry, kept his aircraft flying steady and level with his gyroscopic autopilot invention as far back as 1912.[4] But it took decades more to realize the full potential of the autopilot with the introduction of fly-by-wire aircraft, where the pilot managed the aircraft control surfaces through electronic signals rather than through direct mechanical couples to the actuators. The first pure fly-by-wire aircraft with no mechanical backup was the Apollo Lunar Landing Training Vehicle (LLTV), first flown in 1964.[5] The F-16 was the first production fly-by-wire aircraft.[6] A fully integrated flight management system to control all phases of flight was introduced in 1982: for the first time it allowed machines to execute the full sequence of maneuvers for the duration of a flight, from takeoff to touchdown.[7] So it's no surprise that only a couple of years later we saw the first self-driving car.

In robotics, breakthroughs in one domain inspire similar innovations in another. In 2014, Google Wing completed its first real-world drone delivery, and in 2015 Tesla's autopilot hit the roads.[8] The concepts, sensors, computers, and software are refined piecemeal for different applications but can be used widely, and so end up converging to produce what appears to be sudden technological advancements. If you're surprised when you pass a robot at your local grocery store and worry that the new robot may not be capable of safely traversing the aisles in the midst of small children chasing their parents, you may take some comfort in the fact that the technology inside of it has been progressively refined over the course of at least four decades. How did we reach this moment—where working robots, with the potential to drastically improve our lives, are at once about to become ubiquitous and also too complex for us to control ourselves, leading us to rely on these robots in unprecedented ways?

To better understand how one technological advancement is interlinked with the next, consider our own interrelated systems as intelligent beings: the human sensory system, the network of nerves spread throughout our bodies, and the central nervous system, including the spinal cord and brain. The existence of each component on its own is marvelous, but when put together, it's magical. A sensory system is needed in any complex system, whether it is controlled by a machine or by an operator. Onboard an aircraft, the many instruments include heading, airspeed, and altitude sensors. Sensory systems for terrestrial applications—such as robots in the grocery store or self-driving cars—focus more on imaging the environment and identifying obstacles. These may include cameras, which see things as we do and perform best when lighting and weather conditions cooperate, or radar and Li-DAR, which send out radio and light signals, respectively, and detect ranges based on the signal reflected and bounced back. Radio waves are reflected especially well by materials with electrical conductivity, such as metal, which make it great at detecting standard infrastructure and objects such as vehicles or robots across all lighting and weather conditions. LiDAR illuminates a scene with a laser and measures the reflected light. It reliably detects objects with high accuracy in both day and night, but degrades in poor weather conditions, such as fog and heavy rain. Most robotic solutions use multiple types of these sensors and fuse the data to cover all desired scenarios, just as our eyes, ears, hands, nose, and mouth each tell us different things about the world.

In humans and other animals, the central nervous system is only useful if there are sensors to connect to a central decision-making node. A machine's central nervous system is similar: it is the fly-by-wire or drive-by-wire technology that, much like the human nervous system, transmits electrical signals along nerves to control the body.

Finally, the brain takes the sensory inputs, forms an understanding of the world, makes decisions based on that understanding, and implements actions through the other components of the nervous system. In some animals, the brain can be simple—like the first Roomba, which followed one basic rule: clean in a spiral pattern until it hits a wall, and then follow that wall. In others, it can be complex, like the autonomous vehicles we see on the road that are increasingly capable of navigating construction zones and traffic jams.

We first invested in automation for aviation and industrial applications—such as in the control centers of nuclear power plants—when the tasks required to work these new technologies proved too hard for humans to perform reliably. Today, many safety procedures for nuclear power plants are fully automated and require no human intervention. The systems are too complex for people to monitor or to control manually, and the consequences of a failure are too high. In other words, automation allowed us to reimagine what we were capable of after we had maxed out our natural capabilities.

Robots excel in these applications in aviation, power plants, and factory settings because the environment is tightly controlled and there are detailed task procedures for them to follow. And so robots first infiltrated these industrial worlds that very few people have access to, or were placed in spacecraft, which even fewer people have access to. In these isolated worlds, engineers developed robots' sensors, nervous systems, and brains piece by piece, then tested them, learned from their failures, and improved and hardened the new technologies. Today, by and large, complex industrial applications are heavily controlled by automation with minimal human supervision.

The question now before us concerns the most effective way of using robots to automate our daily lives. We are in the midst of an active debate among manufacturers, legal scholars, and engineers about whether to automate various aspects of driving, and if so, how. Is a hybrid human-car approach the best, or should the person be removed entirely?[9] In fact, we already had to answer these sorts of questions decades ago, when designing the first moon-landing spacecraft, as well as when introducing cockpit automation for air transport. As shown in figure 3, history bears out that seemingly small design decisions can have repercussions for decades, and we'll walk through some examples.

The Apollo Lunar Landing Training Vehicle was the first fly-by-wire aircraft, essentially the first aircraft with a central nervous system. Other advancements brought the Apollo Guidance Computer, which served as the brain of the Apollo Command Module and the Apollo Lunar Module. Landing on the moon was certainly a complex problem,

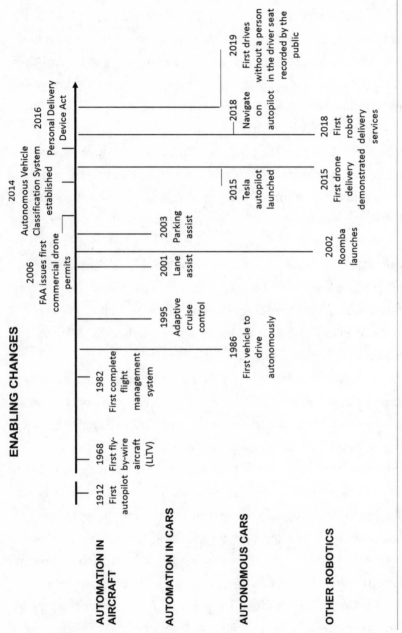

FIGURE 3: Timeline of key enabling changes across robotic applications to depict the technological trends that cross domains and have led to the phase transition that now brings us working robots.

one that neither human nor machine could do alone. The technical debate started then over how much authority the computer should have and how much authority should remain in the hands of the astronauts.

In a seminal paper in the 1960s titled "The Role of Men and Instruments in Control and Guidance Systems for Spacecraft," designers of the Apollo Guidance Computer contemplated just this question.[10] They debated how much to trust the computer in moments of crisis or when time-critical decisions had to be made. The engineers described three types of events based on how decisions are made. Type I events involved foreseeable conditions with predetermined responses, such as the automatic cutoff of a rocket stage at a predetermined velocity. These situations, the engineers said, would be easy to automate. Type II events involved foreseeable situations for which "no appropriate . . . action could be programmed in advance due to the complexity of the general case . . . like landing on an arbitrary spot on the moon." The engineers concluded that Type II events were not straightforward enough to fully automate, but that human performance could be "enhanced with feedback on performance indicators and display of only relevant information." Finally, there were Type III events, defined as those that could not be anticipated, even by the designer or the pilot. The engineers determined that humans would far outstrip automation "in making decisions based on incomplete data in completely new situations."

All of this remains true today, except that technological advancements—in particular machine learning—have changed what we consider "the complexity of the general case" or "a completely new situation." Whereas in the past we had to craft decision-making rules for the automation for a variety of situations by hand, machines can now leverage data or demonstrations to learn an approximation of our human decision-making criteria that may be too complex or time-consuming for us to manually specify. Still, when the unforeseen happens, automation often falls short. In 2009, in what would come to be known as the

"Miracle on the Hudson," the pilots of US Airways Flight 1549 successfully ditched an airplane in New York's Hudson River after the plane struck a flock of Canada geese and lost all engine power. Everyone on board survived. We can nearly fully automate the flying of a commercial airliner, but could an autonomous airliner have pulled this off? It is unlikely with today's technology. No artificial system today can replicate a human's capacity for creative problem-solving.

Even in well-established industries, such as aviation, where automation is prevalent, we still debate what flight deck automation should control and what a pilot should ultimately still control. When you board an airplane, you may not think much about whether it is an Airbus or a Boeing. But decades ago, the two companies chose quite different paths representing different philosophies for how pilots and intelligent automation were to work together. When automation systems were first introduced, if you were riding in an Airbus, the plane's systems would have had authority over the pilot, but the opposite was true on a Boeing. So by and large, the automation could override the pilot in an Airbus, and the pilot could override the automation in a Boeing. These approaches resulted from a basic decision in the design of robotic systems that had to do with hard versus soft automation.[11] Hard automation has more protection against human error, essentially constraining the user from doing something that would put the vehicle in danger. Soft automation still employs safety constraints, but it considers the automation an aid: it provides an alert when the user is about to do something that may be dangerous, but allows the user to proceed and to override the warnings if they choose to do so. The latter allows for more human creative problem-solving. In other words, with soft automation, the user always has access to the full capabilities of the vehicle, whereas with hard automation, there are certain circumstances in which they do not have that access.

Of course, there are pros and cons to each style. For example, in 1985 a China Airlines Boeing 747 had an engine failure while cruising at forty-one thousand feet. It entered an uncontrollable dive and plunged more than thirty thousand feet. With the soft automation system, the pilots recovered control and passengers sustained few injuries. Analysis indicates that a hard automation protection system would have disallowed the pilots' inputs, the very thing that enabled them to successfully regain control of the aircraft. On the other hand, it could be argued that an Airbus flight control system, with its hard automation protections, would not have allowed the aircraft to enter the uncontrolled dive to begin with, because it would have prevented the pilot from taking the actions that got the system into the unstable state. Overall, it seems that preserving the human pilot's capacity for creative judgment, decision-making, and action may have been the winning strategy. An analysis of nine major automation-related accidents from 1988 to 2002 found that Airbus had twice as many accidents as Boeing involving breakdowns in human-automation coordination.[12]

Both of these models face challenges in everyday life. We live in a world full of Type II and Type III events that are far beyond the capability of modern machine learning and AI systems to manage. How do we design these systems in a way that unobtrusively integrates them into our unpredictable world? An awkward robot can be worse than no robot at all: it can become a nuisance, at best, and at worst can create new and unexpected safety risks.

The history of automation follows the history of technology. People have been inventing tools to seamlessly extend their capabilities since the beginning of time. In preparing meals, for example, we started by using fires for cooking, and a rag or other abrasive material to help in cleaning up afterward. But the tools became more complex and turned into appliances, which now relieve people of entire tasks, such as washing dishes, or vastly reduce the time or effort that has to be put into cooking dinner. In some ways, the process of sourcing, storing, preparing, and eating food has become easier; in other ways, it has become more complicated. With automation technology and consumer robotics, these tasks will become both easier and more complicated still. The challenge is to build systems that are not only safe and convenient, but powerful and reliable.

In a sense, our domestic robots are, by and large, fancy appliances, and they still need our attention. We have robotic lawnmowers that do what manual lawnmowers did, for example, but we still need to move them into place, set up a barrier around the yard for them, free them when they get stuck, clean them, and so on. It makes sense that this type of robot would be one of the early successful robot consumer products, because it performs well-defined (aka Type I) tasks that are narrow in scope. These new lawnmowers are similar to industrial robotic systems in that they are programmed for a very specific task and environment. Even small changes require a herculean human effort to reconfigure the technology, because it is programmed with extremely detailed and precise commands that it simply executes. There is no autonomous decision-making: people augment the capabilities of industrial robots today with specific instructions.

In contrast, the new robots transporting us across cities, delivering food to our homes, and keeping our neighborhoods secure are a fundamentally new breed of technology. These working robots operate with an unprecedented degree of autonomy in our complex human world. Aviation and the space program were among the first efforts to grapple with technologies out of the industrial systems mold, in which the automation had to be co-designed with models of human capability to contend with Type II and Type III situations.

The good news is that we have all the tools we need to meet this challenge. We have separately honed the technologies for our robots' sensory systems, nervous systems, and brains, and we have been able to bring together the right combination of technologies for new applications. Similarly, we have a rich set of tools, developed over decades, to fit the capabilities of robots to the limitations of humans, like a jigsaw puzzle. We will be able to leverage our incredible capacity for learning

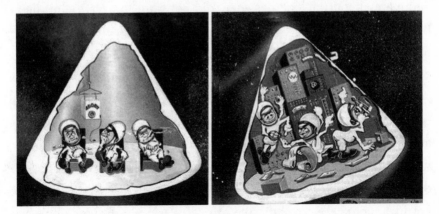

FIGURE 4: This MIT Instrumentation Lab cartoon shows the impact of the extremes of automation. Even during the design of the Apollo Lunar Excursion Module, engineers were thinking about how much to automate and how much control to leave in the hands of the astronauts. If they automated too much, they feared it would leave the astronauts bored and unable to intervene if needed, but if they did not automate enough, giving the astronauts manual control over too many things, they feared they would overwhelm them. *Source:* NASA.

and adapting to solve the robotic challenges of the future, making the adjustments that are needed for working robots to augment our human capabilities and enhance our world. These solutions have been forged across the fields of aerospace engineering, human systems engineering, and cognitive science.

Getting it right will not be easy. You might even call it rocket science. But as we saw with the Apollo Guidance Computer, designing the partnership between robots and people will be fundamental to the success of automation in these applications. Our lunar landings were successful because the design focus went well beyond the underlying technology to include an analysis and understanding of human psychology and decision-making, in order to create systems that would enable people and robots to collaborate seamlessly at the right times and in the right ways.[13]

The cartoon in figure 4, from the 1960s, captures the thinking at the time. NASA needed to find the right balance between automation and manual control. The designers did not want to overwhelm the astronauts with too many tasks during the lunar descent, because they might not be able to keep up and perform adequately. But they also did not want to automate everything, lest the astronauts become disengaged, and fail to intervene if and when their intervention was needed.

We have a similar conundrum with working robots. In chapters 2 and 3, we will discuss hard-earned lessons from aerospace and industrial applications that cast doubt on the vision of a working robot that does everything independently, like Rosie from *The Jetsons*. This robot ideal actually isn't realizable or even desirable. Robots don't have the same capabilities as people, and they don't think like us. This is a strength, but to understand how to fit robots with humans, we first need to understand our own human limitations and strengths—including our propensity to trust inappropriately. The user-robot partnership must be designed from the beginning with this in mind to ensure that these new social entities are effective and responsible. Just because Rosie could flip pancakes doesn't mean you should leave her unattended to put together Thanksgiving dinner.

To complicate matters further, the world that working robots must navigate is exponentially more complex than in a cockpit or factory. Public spaces are busy and change often. New roads are built, sidewalks are closed, storefronts change. What's more, many people will come into contact with a robot who have no idea what it is doing or how it may interfere with their activities. We accommodate people we don't know every day as we pass them on the street or at the grocery store, but this kind of task is very challenging for a robot. In chapter 4, we will discuss how our robots will have to be able to develop at least a minimal awareness of bystanders and accommodate them as they move along.

Of course, designing a lunar lander is quite different from designing a consumer product. There will be lay users instead of highly trained astronauts involved, and there has to be a positive business model. In fact, as we will see in chapter 5, the commercial world is not well positioned to guide the crossover of robots to society today. The way companies design products and what consumers seem to want from them are at odds with how autonomous social systems will have to be designed. Working robots are typically able to perform particular tasks with little or no interaction with people. Consumer products are designed essentially to delight and entertain their users. Moreover, there is little or no focus today on how the design of a product affects the user's ability to perform other tasks effectively. Does Facebook care about the impact on your work productivity? Consumer products like social media platforms are designed to be fun, which often comes at the expense of being productive, transparent, and robust. And they can afford that trade-off, because the stakes involved in our interactions with them are fairly low. If Twitter goes down, nobody gets hurt. Working robots, in contrast, are there for us to offload tasks to them that *must* be done, and that people do not want to do or cannot complete safely on their own. If those robots don't work well, or distract us at exactly the wrong time, there could be material consequences. And the design model must change to focus on the best approach to ensuring successful execution of the robot's task.

In fact, research has shown that a user's preference is often in direct conflict with the design that leads to the best performance, or even safe operation. Users are often drawn to and enjoy systems that are not the most helpful to them in the task at hand. In other words, features that delight users typically are not optimized for the best performance result.[14] For working robots, this means that a user's intervention may be needed when the user wishes not to be bothered, or that the user may not be needed when he or she would like to participate. Simply put, working robots that delight users may not actually perform the best. And this matters even more now than in the past because the stakes are higher with working robots than with any consumer product of the past. Robots are powerful machines that are highly capable, but they also pose a risk to society if they are not safely introduced and managed.

Our world is complex and dynamic, and the creativity and judgment that come so naturally to us still elude even the most advanced robots. Robot brains are still simple, doing exactly what they are instructed, programmed, or trained to do. Our only chance for successfully integrating working robots into our everyday lives is to embrace this fact and reprogram our consumer product design philosophy to focus on building the right partnerships across robots and people.

For working robots to be successful, we need a new language and approach for them to communicate with people and with each other. In chapter 6 we will outline the need for automation affordances so that when you engage with a working robot, you know what to do or say to influence that robot without requiring special training or expertise. And these automation affordances must be clear enough that a user can rapidly understand how and when to take an action, since the robot may pose a safety risk for an outcome that only a person can prevent. In chapter 7, we will describe the need for robots to communicate with each other to resolve direct conflicts, coordinate activities, and share knowledge about the world that may give all robots a better chance to succeed, similar to crowdsourced mapping products we use today. These mapping products help us navigate around cities and avoid traffic jams, but they are even more important to robots, which have more disadvantaged sensory systems and are not as flexible as humans when reacting to the unexpected.

Not only are the users of new working robots not experts, but their environments are chaotic. We can't simply fence off space for working robots if we want them to take on tasks like pizza delivery. In chapter 8 we will address the design of the environment for working robots in daily life, which—from a robot's perspective—is uncontrolled and inconsistent. Think of the simple challenge of staying within lane lines on the streets of an urban setting. Construction projects, weather, and normal roadway wear and tear hinder our best attempts at providing clear lane designations. People are very good at adapting to these variations and are extra vigilant when the structures they depend on degrade. But robots are not yet capable of navigating such ambiguities. Even minor variations in the physical environment are very disorienting to a robot—and they make the design of safe robots very challenging, if not impossible, at scale. If we cannot rely on consistent lane lines, then the robot designer must come up with a different way of determining how to stay within a lane. The autonomous driving community is solving this problem by developing very precise maps, on the order of centimeters, for every road they drive on, and then incorporating very precise localization solutions to be able to know the robot's position extremely accurately within that map. This approach requires high-end sensors. All of this is expensive to get right, and even then, some vulnerabilities remain. Not even the most exquisite machine can solve the problem—rather, it will require the economic and political will of society to make changes to both private and public infrastructure.

If we can embrace this reality and start designing working robots in a new way, then we have a shot at improving our daily lives with new forms of dynamic appliances. These new appliances will rise off of our countertops and out of our homes to start helping us with all sorts of tasks, and in the process, they will capture and use data that wasn't possible for them to capture before.

As working robots scale, that data will become more broadly available, and we will have an opportunity to collect and share it to improve the performance of robots across company boundaries. In chapter 9, we will describe how this kind of data sharing has been useful in aviation and how the lessons from aviation can be applied to robotics. In aviation, for example, there is an incidents and accidents database that anonymously gathers data on near-misses and accidents across airlines, airliners, and pilots. This database provides a foundation for tremendous knowledge on the safety and vulnerabilities of automation in the skies. Data gathering and sharing are even more important with working robots because they are developed and trained by volumes of data.

While companies may prefer to maintain their competitive advantage by hoarding data, society has much more to gain if we can unlock these treasure troves for everyone's use.

Our hope is that this book will help us as a society to rethink how we design working robots so that they can become responsible social entities, and so that we will be able to make adjustments to our world to accommodate these new entities, which have unique needs and present new challenges. Some of the solutions we recommend can start the conversation about the critical intersections between technology and society, so that we may realize this future where working robots begin to help us in our everyday lives. It is our moment to take the capable tools we have developed and take the next leap. Should we choose to accept this mission, we will be ready for working robots.

CHAPTER TWO

There Is No Such Thing as a Self-Reliant Robot

IT'S DUSK, AND YOU'RE DRIVING HOME FROM WORK ALONG A hilly road. It's windy. You didn't get enough sleep last night, and you have a headache. So you turn on your advanced driver assistance system for some help. The car is driving automatically now, staying in the lane and keeping a safe distance from other cars. You breathe a sigh of relief and start to relax.

As you round a bend, the glow of headlights indicates that another car is about to round the bend toward you. Suddenly, there's a "ding," and you begin to veer out of your lane. You stiffen, grab the wheel, and turn the car back into your lane. If you hadn't, you would have gone careening into the ditch. What happened? Why did the system designed to assist you turn off? What if, in that moment, you had been looking away, perhaps at your phone? You shudder just thinking about it. The next day, you file a complaint with the car company, and eventually a software patch is pushed out to your model. It was an "edge case," a seemingly unlikely situation that wasn't anticipated during the manufacturer's simulation and testing—your car didn't understand what it was sensing, so it gave control back to you. The engineer had decided that, in edge cases, a person would know what to do better than a robot.

When we imagine the future, we like to imagine robots that are capable of completely taking over routine tasks. And engineers attempt to design robots that mitigate as many types of failures as they can identify as possibilities. As shown in figure 5, several layers of protection against failure are being built into modern-day intelligent systems. But the first step to making safe, powerful working robots is to recognize that completely self-reliant robots are a fantasy. Failure is a part of life, not only for humans, but for robots, too, and no amount of planning and testing can change that. It is not reasonable to expect to be able to identify or account for all possible error circumstances that might arise in day-to-day living. The human world is simply too complicated to predict.

And even if a robot's software doesn't fail, its hardware still will, at some point. So it's neither feasible nor responsible to try to design working robots—which will be in charge of safety-critical tasks—to be truly self-reliant. Just as humans work best when they work together, robots need to be designed, above all, to be good partners: good partners to the people who work with them, to anyone else who comes into contact with them, and to the society in which they operate. Being a good partner is as much as anything about knowing your limitations and how to plan for them. Design approaches of the past have not tapped into the full potential of a human-robot partnership to help robots when they encounter the unanticipated or face rare failures.

Robots can fail in lots of ways. Their physical components can break. Their sensors can fail. The power can go out. The system may encounter an unforeseen and unplanned-for situation. Some robotic systems—such as autonomous cars—operate with software so complex that it's often not mathematically possible to enumerate all the situa-

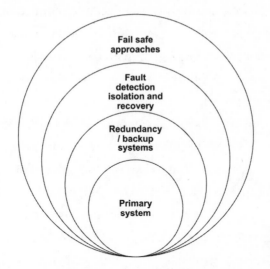

FIGURE 5: Traditionally, automated systems have been designed to mitigate and overcome known errors through layers of protection, such as redundancy in case a primary system has an electromechanical failure, fault detection isolation and recovery when errors are identified, and fail-safe approaches to try to ensure human safety even when there is a catastrophic system failure.

tions one would need to verify in advance to ensure that the system will behave correctly.[1] Our human world is impossibly complex. Consider a sidewalk delivery robot traversing through a town center. It might lurch forward to indicate its intention to maneuver around you, but this aggressive stance may be unacceptable to a parent pushing a baby in a stroller. What about when a group of parents with strollers are coming up the sidewalk engrossed in conversation, or if a couple of people have their heads down because they are looking at their phones? What about when it is raining—and the robot can't move to the side because it would run into a deep puddle—but people are pushing past to get out of the rain quickly? Or when two dogs cross paths, sniff each other, and wind their leashes around the robot, while the dog owners dance around the robot trying to untangle it?

If the robot's program had just five conditional decision points, such as checking for and responding appropriately to the five previous situations, and it had to continually collect a series of twenty data points to identify the current situation and appropriate response during its traverse, the program would have 10^{14} possible paths through the code. If the designer tested one of these paths a millisecond, it would take over three thousand years to fully test the program. And this is just one robot, doing one job, in one specific location. The potential of collaborative robots lies in the ability of a machine to work successfully in a variety of contexts. Exhaustive testing is not possible, so we must plan for the system to sometimes behave in ways the designers did not expect.

Of course, one way to minimize complexity is to remove human interaction whenever possible. People, and the environments they build, are the ultimate uncontrollable entities. But this idea negates the entire point of creating collaborative robots. A system roaming our offices, hospitals, and streets will necessarily interact with people. We make robots in the first place to make human lives easier.

We may design a system to minimize reliance on a person because there are some situations that unfold too quickly or include too many components for a human to be able to effectively check a robot's actions. In these situations a "high level" of automation is needed. Designing humans out of a decision requires special considerations, which we will explore later in the book. But most of the time, the robot does not have to be truly self-reliant, and to be fair, neither do people. People make mistakes, get confused, and ask for help from others all the time. We seek out a local in the city to give us directions, or ask for help in getting a subway ticket from the information desk. Most of human life is facilitated by the help we give and receive. Robots should be able to seek out or rely on help from people too.

In order to do that, robots first need to know when the help they're receiving is actually helpful. Especially when a human is required for operation of the system, the system must be specially designed to guard against errors that a faulty human might introduce.

To tackle that problem, we first ought to come to terms with what human error really is. A human error is an act (or omitted act) that was "not intended by the actor; not desired by a set of rules or an external observer; or that led the task or system outside its acceptable limits."[2] Human error may cause a system failure and could stem from a number of possible root causes, including specific acts or omissions, such as forgetting to perform a step on a checklist. It can also be the result of preconditions for specific acts not being met (such as driving while tired or impaired); inadequate supervision (for example, an unlicensed, inexperienced driver getting behind the wheel of a car); and organizational influences (such as differences in the rules of the road from state to state, or country to country).

Failures can be active or latent. Specific unsafe acts or omissions that lead directly to an error, such as skipping a step on a checklist, are called *active failures*. However, *latent failures*, including those stemming from organizational structure, lack of supervision, or neglect of preconditions, are often more insidious. They may lie dormant for days, weeks, or months until they contribute to an accident. Say you work the late shift and drive home night after night exhausted. What is there to do about it? Each night you get home safely, until one unlucky night you forget to check for bicycles when you start to make a turn on red. Because latent errors are things we don't pay attention to, they are almost by definition harder to identify than active failures, and addressing them may require substantial efforts that go beyond the design of technology.

How do we shore up our robot teammates, to know when and how they need help? As our starting point, we can look to how designers guard against failures due to human error in industrial systems. The key idea is robustness across layers: industrial designers recognize that any one layer of oversight could include unforeseen weaknesses, so they design systems with many layers of defensive protection. Industrial systems designers use a "Swiss cheese model" of accident causation to depict the emergence of an accident as a failure across many layers of defense.[3] There are flaws, or holes, in each layer that, if aligned, create a vulnerability in the overall system. Plugging the holes to avoid specific unsafe acts is one strategy for safeguarding against the inevitable errors that humans introduce. However, it's difficult to anticipate every hole and close them up entirely, even with the billions of dollars of investment that go into, say, the development of a new commercial airliner. For working robots, the success of that approach alone will be quite limited, because it's not possible to root out every possible gap. The more layers of protection you can engineer into the system, the more effective your system will be, because by having these layers you minimize the chance of ending up with matching holes. Neither robots nor humans alone can be error-free, but we believe that the partnership of a human and a machine *can* be error-free, or at least come as close to error-free as one could hope, especially because robots and humans will likely have different kinds of defenses against failure, making up for each other's shortcomings. The results of such an approach can be seen, again, in aviation, where the human-machine partnership has led to an unimaginably good safety record, given the number of flights that occur daily. If we can achieve such gains with human-robot partnerships in more earthbound applications, we will see incredible improvements in the safety of our streets and sidewalks.

This Swiss cheese model can be used in designs for working robots. The foundation of the model is the human-robot partnership. But what does a partnership look like? In any good partnership, each party watches out for each other party, and is able and willing to intervene when one of them needs help. To be ready to intervene, a teammate must understand what that partner is trying to do and how it may be tripped up. In human teams, a partner who can understand her team-

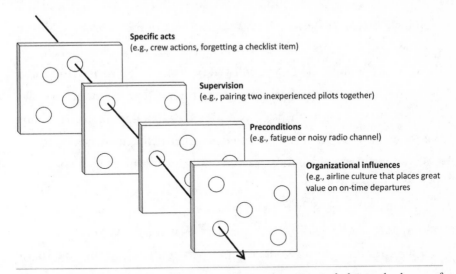

Specific acts
(e.g., crew actions, forgetting a checklist item)

Supervision
(e.g., pairing two inexperienced pilots together)

Preconditions
(e.g., fatigue or noisy radio channel)

Organizational influences
(e.g., airline culture that places great
value on on-time departures

FIGURE 6: This diagram shows how circumstances can create holes in the layers of protection against error to expose just the right situation for an error to occur. *Source:* Adapted from James Reason, "The Contribution of Latent Human Failures to the Breakdown of Complex Systems," *Philosophical Transactions of the Royal Society B* 327 (1990): 475–484.

mates' thinking and capabilities can better anticipate their responses, and then can better communicate with them or act to guide them in ways that will help them avoid or mitigate error. Likewise, our new working robots need to be able to understand the potential weaknesses of their human partners and must be built to be able to shore up those weaknesses. The user also must understand, at some level of detail, what the robot is trying to do, and must be able to identify situations that may flummox the robot in real time as they arise. When the user is an expert and receives thousands of hours of training, this part is easy. In that case, the user understands how the system works in detail and has developed a rich mental model to consult. Even then, the task can be very hard. Everyday consumers will have much less training time when working robots start showing up in their lives, and therefore a shallower understanding of the systems, their operation, and their weaknesses.

A pedestrian, for example, can't be expected to understand the technical details of how a sidewalk delivery robot's sensors scan the environment to create a three-dimensional textured map of the world— that is, a 3D point cloud. Nor will the pedestrian know how the robot processes the point cloud to understand the objects around it, in order to select its walking path. We need a new design approach that trains the robot to understand its teammates—that is, in this case, pedestrians on the street—since those people in its environment won't have the luxury of training.

The good news is that having a shallow understanding of a robotic system may not be as difficult a problem as it seems. We often assume that in order for a person to learn, an instructor needs to be able to assess what the student already knows. If someone has already made an error, what misconceptions led to the error? It is only by diagnosing the misconception that we can fix the problem at its root. Even better is to be able to identify those misconceptions in advance, to avoid the error altogether. Educators call theories about this problem the *diagnostic remediation hypothesis*. And although this area of research may apply to the question at hand, diagnostic remediation feels like an impossible dream for the human-robot partnership.

Even in the area of human tutoring and instruction, the set of lessons we have learned can seem counterintuitive. There is evidence, for example, that human teachers do not always employ diagnostic remediation. Studies indicate that they rarely utilize knowledge about students' misconceptions, even when this information is explicitly presented to them.[4] Nor does trying to explicitly address students' misconceptions always help them to learn more effectively—at least in the case of algebra, as one study showed.[5] One reason is that misconceptions do not always result in systematic errors, and therefore fixing a misconception does not always have the expected effect on performance.

This area of research gives us hope that robots do not need to know everything about users' thought process and biases in order to effectively ask for help. Rather, given a situation that involves human intervention, the robot only has to determine what it needs its teammate to do to overcome the issue. And a general model of how to effectively influence a robot teammate is sufficient for guiding a user toward the correct sequence of decisions or actions needed to solve the problem at hand.[6]

Additionally, working robots and their users will not be operating within highly structured organizations that maintain strong systems of oversight, such as the airlines, and robotics system designers will have to account for that fact as they consider how to avoid Swiss cheese–style holes. For consumer products, societal influences—such as the regulatory environment, cultural norms, and infrastructure—will affect attitudes toward robots, the environmental support (or lack thereof) for improved system performance, and additional opportunities for both latent errors and protective defenses. For example, the design of sidewalks and roads can either help robots or make their jobs harder. Patchy or missing lane lines make it harder for a self-driving car to follow the road consistently. A construction zone marked with scattered cones and no clear way around it is very hard for a robot to understand and navigate.

Cultural norms are another factor in human-robot partnerships. Just as these norms affect relationships between humans, they also affect relationships between humans and their robot helpers. In aviation, for example, it's been found that cultural norms influence the interactions among pilots, other crew members, and cockpit automation features. These norms are in turn influenced by other factors, including a person's national culture, the professional culture of the groups to which the pilot and crew members belong, and the specific organizational culture for an airline.[7] Different cultures often have different attitudes toward power distances depending on whether they are more individualistic or more collectivistic. These differences influence the way people collaborate and the cultural barriers for effective collaboration during critical moments. In high power distance cultures, the relationship between the team leader (the pilot) and his or her team members (the copilot and crew) is less equal, with a deference toward the leader (the pilot). This manifests in an unwillingness for team members to question or challenge the decision of a leader and the leader's unwillingness to listen to points made by subordinates. In lower power distance cultures, such as the United States, all crew members are expected to participate in solving problems and to uniformly volunteer pertinent information without concern about their power state. High power difference can contribute to accidents when lower-status members of a team withhold critical information from their leader because it conflicts with what they think their superiors want to hear, or when they do provide the critical information, but the information they share is ignored.[8]

The impact of cultural norms on human-robot interaction is somewhat similar. Nonverbal behavior in human-human interaction varies across cultures, for example. The norms around how much personal space is maintained between people when they are interacting is called *proxemics*. Researchers from Cornell University found that the cultural differences in proxemic behavior extended to human-robot interaction: in cultures where people prefer to be closer when talking with someone, they will expect the same proximity with a robot, and in cultures where people are sensitive about their personal space being breached during a conversation, they will expect robots to also respect that personal space.[9] This simple example is just one way that we can expect culture to impact the relationships between robots and the humans who depend on them.

To make matters even more complicated, robots will increasingly be interacting with bystanders who are fundamentally unpredictable, and there will be only a limited number of ways for them to communicate with each other and influence each other's behavior. The broad and unpredictable set of situations that may arise could cause latent errors to surface.

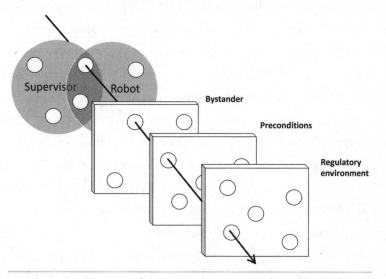

Figure 7: An adaptation of the Swiss cheese model for working robots in commercial applications.

While in some ways it may be harder to fill in the holes of the Swiss cheese slices for these kinds of preconditions and societal influences, the human-robot partnership does offer opportunities for designers to cover overlapping blind spots. Most importantly, perhaps, is that both robots and humans will be able to identify and overcome the potential errors of the other. Thus, figure 7 shows the human-robot partnership as interlocking circles, with overlapping protection as well as exposed areas for each. The stronger the design of the partnership, the smaller the exposed areas will be.

In this model, the robot not only helps to prevent human error, but the human user has an important role to play in mitigating the robot's errors. For a person to be a good partner to the machine, she has to be capable of predicting failures that the system doesn't know about or can't respond to. However, a system that blindly relies on a human "safety net" to catch and fix errors is doomed to failure.

For example, in the six levels of automated driving features defined by SAE International, a professional association governing standards in engineering, there are three levels on which the driver is no longer in control of the driving feature. But for Level 3, the human driver is still expected to be able to take control of the vehicle when the automated driving feature requests help. On Levels 4 and 5, the automation does not require the driver to take control; instead, the responsibility is on the designers of the automation to determine failure detection and recovery approaches other than relying on driver monitoring.[10]

Level 3 automation is challenging because the automation takes over much of the traditional role of the human driver, but to be effective partners, drivers need to know how and when the autonomous drive mode might fail, so they will be ready if and when they are needed. For example, the car is likely to need human intervention when it encounters unexpected obstacles, such as roadwork or detours. Ideally, the person is aware of this possibility and is on constant lookout, so that he or she can attend to the situation immediately when it arises. A fully autonomous vehicle with Level 4 or 5 automation needs another layer of protection. The autonomous car's software must be able to detect when it is outside the limits of its capability. When these situations arise, the robot must put itself into a fail-safe state, such as coming to a stop when approaching an array of scattered traffic cones. There is still a human supervisor, in addition to, perhaps, a passenger, but they are in a sort of "command center," and their help is requested only when the autonomous car detects a situation that it can't handle itself. This is still a human-robot partnership, but there is even more responsibility on the robot to help the supervisor help it.

We can design this kind of partnership in much the same way that we bring together people to make high-performing sports teams. A key part of the success of a modern-day sports team is paying attention to how teammates complement each other's skills and expertise. Use of data analytics and statistical analysis, popularized for baseball in the book *Moneyball*, can take a lot of the guesswork out of the process. In baseball, this kind of statistical analysis is called *sabermetrics*, a term derived from the acronym for the Society for American Baseball Research. We can do the same with human and robot teams. For example, we can develop metrics analyzing aspects of human capability and fallibility as well as machine performance as they pertain to the human-robot relationship. As proof of concept, our lab at MIT developed a machine-learning model that allowed a robot to observe a human performing a task imperfectly, and then make a guess about what the person didn't see or wasn't paying attention to; it also attempted to determine which aspects of the human's imperfect behavior might be due to not understanding the situation fully, versus which aspects could be due to decision-making mistakes about the appropriate action to take, despite understanding the problem.[11] For example, if a health-care worker gives a patient the wrong medicine, might it be because of similar labeling between two vials, which results in the health-care worker picking up the wrong vial some of the time, or because of deficiencies in the worker's training about the appropriate medication to use? A robot that can model our blind spots—our human holes in the Swiss cheese model—can better catch these errors and shore up our performance.

Furthermore, we also know that a team of experts does not necessarily make for an expert team.[12] People learn to work in partnership not only by virtue of their vast knowledge, but through experience and practice working together. We use some previous knowledge about our teammates (their skill sets, prior work experiences) and about the task at hand (skills or procedures), but ultimately, we need to practice working together with team members to "get in a groove." We develop a familiarity with each other, learn each other's idiosyncrasies, together patch the "holes" to anticipate and avoid failures, and learn to trust one another.

Which baseball team do you think might perform better? A "dream team" of all-stars, playing together for the first time, or an average team that has spent the season training together, but now, on game day, has to play with their positions scrambled—the shortstop has to pitch, the first baseman has to play outfield, and so on? You might guess the dream team, but you would be wrong.

A human factors expert, Nancy Cooke, and a sports psychologist, Rob Gray, teamed up to conduct just this study.[13] They formed three baseball teams—one that played with each member in his typical position, one where the players swapped positions, and one all-star team made up of the top performers for each position across a number of different teams, all playing together in their typical roles for the first time. The researchers recorded the teams playing games, and then tested each team's performance. They showed each team member video clips of pivotal moments of the game—such as right after a hit—then froze the video and asked the team member to predict what each member of the team would do next. The ability to get this right provided a measure of the team's cohesion. The team that practiced together and played together in their typical positions performed best at this task. But interestingly, the "dream team" performed miserably compared to the team that played in scrambled positions. The learned familiarity with their team and learned model of how to work together were more important than any skill players had developed in a single position.

Building up an understanding of how to work with one another prepares a team to be better at problem-solving. Their comfort in working together makes them more flexible and agile to jointly respond to situations that arise. Psychologists have devoted decades of research to understanding and designing team training procedures that optimize a team's ability to respond to unforeseen events.[14] What they have found is that teams that dedicate practice time to responding to many different sorts of errors—a process called *perturbation training*—get better at responding to new types of errors they haven't yet seen. In other words, given the right selection of error scenarios, teams learn rhythms and strategies for effectively responding to any new error situation. This is similar to the baseball team that has practiced together and then has their roles completely changed. They are able to adapt to new situations.

We are only beginning to understand how to design robot team-mates that emulate the special ability of human teams to respond to unforeseen circumstances. In our lab at MIT, we have designed new ways for humans and robots to learn together through perturbation training.[15] When a person and a robot practice together on variants of a task, the robot uses a machine-learning algorithm inspired by human cognitive models for learning. It learns to draw from its experiences and refine its strategies for teamwork in ways that are similar to how people learn from prior experiences. In our laboratory studies, we pro-vided teams of people and robots with puzzles to solve—for example, they might have to collaboratively allocate firefighting resources to ex-tinguish a set of fires on a city block in windy conditions. We found that a robot teammate, using human-inspired learning techniques, did in fact learn teamwork strategies that were compatible with the human partner's learned strategies, and that human-robot perturbation training led to a higher level of team performance than could be achieved with alternate procedures.

This is an encouraging sign that we can develop robot partners that learn to work with people in professional settings, such as hospitals, where people are willing and able to devote substantial time to training in simulators. But a person who has just ordered a package that will be delivered to their door by a delivery drone will not be a trained pilot or spend time learning how drones work and when they may fail. And so we need other tools and approaches to designing a more perfect part-nership with working robots that interact with everyday users.

Learning how to use any technology, even very simple systems and interfaces like TV remotes, requires practice. Most of the time, we use trial and error to build our understanding of how a device works. Design processes for "learnability" recognize this aspect of hu-man nature. Rather than trying to design a system so that users can do a task right the very first time, the goal is to ensure that users have clear and consistent cues and feedback, so they can better remember what worked and repeat that behavior the next time they interact with the system. Conversely, user interface (UI) designers take it as a given that users will explore the interface, and so are careful to bury critical functions deep in menus, with many confirmations, to avoid accidental activation of irreversible actions, such as the factory reset.

How do we help the user quickly develop the right familiarity with a robot to be able to identify and overcome robot errors or failures? In safety-critical systems, learning through trial and error can be dangerous and unacceptably time-consuming. So what is our other way forward? The first step is to actively change the task of the user from the tasks they did before they had a robot partner. Basketball players change their strategy when they are playing a one-on-one game instead of on a team. Similarly, people accustomed to working alone will need to change how they behave to achieve success in working with a robot teammate. The users' tasks need to shift so that they focus on watching out for holes, rather than directly controlling the robot. This new task involves monitoring for situations where the robot may fail, so that the user can intervene and help the robot sidestep the hole to be successful.

The system design accordingly will need to focus on how it can best support users in adapting to their new role. For example, when driving a car, traditionally you are constantly scanning the environment in front of you, taking in many inputs simultaneously to drive safely. When a car begins to drive itself using a lane-following technique, the driver should be prompted to shift his or her focus to a general monitoring task—for example, watching out for lane lines that may be obscured in some way, such as in a construction zone or in unexpected lane shifting. A data visualization could be provided, perhaps on the windshield or as a display in the car, to highlight the lane line interpretation, further helping the driver monitor the car's ability to sense and understand the lane lines. But this is just a hypothetical. It would only be appropriate if the developers found that it made sense for passengers in self-driving cars to watch out for unexpected discrepancies in lane lines. It is the job of the robot designers to figure out what the key tasks are for the users and then determine how to support them in these new tasks. This responsibility becomes even more important as robots become more complex and take on more authority.

Once these new tasks are identified, they can be accounted for in the design of the robot in ways that will positively influence human behavior in the interactions that take place. The way data is visualized, for example, influences how people comprehend information and make decisions based on it. Proven data visualization techniques can help a user see the robot's holes and figure out how to maneuver around them or patch them. Bar graphs, for example, have been shown to best support comparisons between discrete data points, whereas line graphs have been shown to best support trend information.[16] How the information is presented will fundamentally determine how the user sees what needs to be done.

To understand how this concept could work in a complex and dynamic environment, we again turn to aviation. Flight management systems support pilot tasks and decision-making, from route planning to the execution of emergency checklists. Studies have found that the way information is conveyed from a flight management system to the pilot can influence the pilot's decision-making.[17] This research suggests that information representation would likely influence human behavior and decision-making with other complex and dynamic systems as well. In other words, if it works for pilots, it should work with the driver behind the wheel of an autonomous car.

A specific design approach that has been developed for this exact purpose is called *ecological interface design* (EID).[18] EID was developed for industrial applications such as nuclear power plant control panels, medical devices, and cockpits. It changed the focus of the design effort from one that simply reduces the complexity of the user interface to one that focuses on revealing the true constraints of the system to support the user in building the right active mental model. Thus, it allows the user to transition from being a passive monitor, which we know people are consistently terrible at, to becoming an active problem solver. The key to this approach is to make the constraints and complexities of the environment perceptually evident directly through the user interface, which allows the user to more directly identify failures and adapt to those failures.

For a simple example of the power of an ecologically focused display, consider navigation systems. Before we had navigation systems in our cars or on our phones, many of us looked up directions on the Internet before traveling somewhere new, and either printed the list of turn-by-turn directions or printed the map with the route overlaid on top. The turn-by-turn directions are easier to follow when things are going as expected, and many of us started with this list. With the map, you had to constantly reorient yourself, hold it up close to read the next street, and so on. However, as soon as something unexpected happened, the list of turn-by-turn directions was useless. Let's say there is construction along your route and you have to make a detour. Without the map, you will not know how to get back to one of the streets listed in your directions, or even if you do, you won't know where you are in relation to your next turn.

While the printed map with the route overlaid on it is harder to use during normal situations, because you have to examine it closely and figure out how it relates to your current location and the direction in which you are traveling, you have to have it on hand in order to recover from unexpected circumstances. We will have to make some of these trade-offs when designing robots, not only to enable the user to formulate an appropriate active mental model of their robot partner during nominal conditions, but also to enable the user to adapt in the face of unexpected failures or the sudden appearance of scenarios that the robot may be unable to predict. We will need user interfaces that make the constraints and complexities of the robot's model of the world perceptually evident to the user. For a robot, this could include a display that shows what the robot sees, to enable a user to easily visualize and detect when the robot is missing key information that may lead to an error. Rather than simply telling users when they are needed—and have to take over from the autonomous system—users should be able to see when the robot is failing to properly understand their shared environment.

Many studies have assessed ecologically oriented displays over alternate approaches in industrial applications, and they have consistently shown that users diagnose problems more quickly and accurately with ecological interfaces than with other types. With ecological interfaces, operators tend to be more flexible and to have a better understanding of the system they are working with.[19]

In EID design, it is recognized that the robot partner is intelligent, but in a different way than we are. It's important to make the following perceptually evident to the user:

- The robot's goals and intentions, and how they relate to the environment
- The constraints on the robot's decision-making and actions
- Indicators of robot performance as they relate to the environment

With this type of design, users are more likely to step in and help as needed, rather than always waiting to be asked. Users are more likely to be able to identify when the robot's goals, intentions, or actions are erroneous, and to quickly address these errors. EID helps users build an accurate mental model of the robot's decision-making capabilities during normal situations, effectively training them on how the system works in the event of trouble. Ecologically focused displays are not needed under normal circumstances, and EID may feel excessive during those times. But unknown errors will inevitably surface, and EID will be critical when a robot's failure creates a problem that a human operator has to detect and solve.

So far in our discussion, guarding against error has meant helping a person observe certain metrics and predict where and when a robot is likely to fail. But it is worth trying to figure out how to prevent errors in the first place as much as we can. Certainly, prevention seems preferable to fixing an error once it's happened.

People are much better than robots at reading the signs and understanding how changes to their environment or situation will impact them. Indeed, robots are often not capable of understanding some of these factors at all. For example, a construction sign may introduce new rules to highway traffic, such as a reduced speed limit, or there may be cones to block off a lane that a robot doesn't know about. A person is highly capable of taking in that information, even if the clues are sparse, and then understanding it quickly and responding, whereas a robot has to be given detailed, explicit information and procedures to follow.

For emaxlpe, it deson't mttaer in waht oredr the ltteers in a wrod aepapr. The olny iprmoatnt tihng is taht the frist and lsat ltteers are in the rghit pcales. The rset can be a toatl mses, and you can sitll raed it wouthit a pobelrm. That you can easily understand those three sentences is a testament to our exceptional ability as humans to fill in the gaps when information is sparse. We are able to recognize patterns and use context to understand the passage.[20] Neuroscientists have discovered that we use context to preactivate areas of our brains that correspond to what we expect to come next. Machines today do not possess this same ability to identify and connect patterns based on general context. A machine would need to attempt to spell-correct every word in the sentences appearing at the beginning of this paragraph. If even one or two words were not corrected accurately, the machine could draw a completely incorrect conclusion, or no conclusion at all.

A person can help a robot in the face of this kind of unexpected and sparse information—information that changes the rules, that leaves out important details that the robot needs, or that can be misunderstood in other ways that may cause failure. If the user is supported in looking out for these situations, and actively solves the problems as they arise, they can communicate with the robot to correct its decisions and procedures. It might mean, for example, that a user will share a new rule with the robot, or inform it about an environmental change. This could be a reduced speed in a construction zone, a pothole, or someone who needs extra time to cross the street. These kinds of obstacles are easy for people to detect, but hard for robots. People understand immediately how to adjust their behavior to accommodate such situations, but robots need to have a specific rule designed to cope with them. Just as people help others by sharing information on apps such as Waze, we can envision robot supervisors, bystanders, and even other robots supporting robots with crowdsourced information to help them find their best routes.

THERE IS POWER IN THE HUMAN-ROBOT PARTNERSHIP BECAUSE THE ROBOT has strengths that the user doesn't have and vice versa. The robot knows some things about the world, but doesn't know everything the human does. For example, robots often struggle to learn nuances of social interaction—the subtle body language, such as a head tilt, to indicate that a person is bored or confused. Similarly, people know many things about the world, but don't know everything the robot does. A robot that assists a surgeon in the operating room never blinks, never fatigues, knows exactly how long a cut is, and doesn't lose track of the location of the surgical instruments in the operating field.

How can we leverage the robot's relative strengths and design interactions that shore up human weaknesses?

First, given an understanding of a human's potential weaknesses and blind spots (i.e., the holes in the human slice of Swiss cheese), we can design the robot's tasks and capabilities (e.g., sensors, behavioral intelligence) to ensure that its weaknesses and blind spots (its holes) do not overlap with the human's. According to one study, nurses and surgeons miscount surgical instruments and sponges in up to 12.5 percent of surgeries, leading to dangerous errors with objects being left inside of surgical patients.[21] Given this knowledge, surgical machines and robots can be fitted with special sensors to aid doctors in tracking surgical items.

Some overlapping holes will still exist. But if we can give the robot a dynamic model of human weaknesses related to the task it is designed to accomplish, then the robot can use this model to actively anticipate when an act might pass through one of the human's holes. The robot can potentially catch the error before it occurs, and direct the human's attention or behavior accordingly (that is, temporarily "patching the hole"). This is similar to how we might put up a mirror to provide visibility around a blind curve, or put up a street sign that says "Slow, Children Playing," in a neighborhood with a lot of kids. These are aids to cue drivers to adapt their behavior and drive more cautiously in these potentially troublesome situations. We can do the same for people when they interact with robots. Although people may have individual weaknesses, we also demonstrate some common gaps in our human awareness and decision-making, and designers can use these patterns to figure out where and how users may struggle. Through experimentation during product design, we can home in on additional areas where the robot will need to guide the user away from the "holes." Or else the robot can identify where a human action has fallen through a hole, and either take action itself to mitigate the consequences or alert the human to the problem.

Perhaps most importantly, the robot can help to prevent or overcome human error by managing the user's attention. People have systematic biases in where they direct their attention, and when the attention is focused on the wrong details, important information can be missed.[22] The robot can be designed to help the supervisor overcome these common human limitations.

For example, during times of stress, a user's focus narrows. This is called *spotlight attention*, and although it allows us to focus intently on a given task or display, it can often cause us to miss information that is on the periphery. A robot might be able to detect these times of stress, and either attempt to redirect the user's attention or autonomously take over tasks the supervisor is unable to perform because it requires too much multitasking.

Additionally, people often perform best in the face of unforeseen events when they are guided toward the appropriate response. When they are provided with too much information about a situation, it can become confusing. In fact, sometimes the more people know about how a system works, the *less* able they are to figure out how to intervene to avoid errors. It's a bit counterintuitive, but sometimes too much information can be harmful and distracting. When possible, robots can help guide users toward the steps needed to address failure.

In the 1980s, researchers conducted studies in which they aimed to train people to operate a simple control panel consisting of a number of buttons, switches, and indicator lights.[23] Participants were told a story about the purpose of the control panel—that it was the control panel for the "Taser bank" on the Starship *Enterprise*. The purpose of the study was to understand what sort of information people needed to know about a complex system in order to operate it correctly. For example, did they really need to know anything about its functioning? Or would it be enough to just train people on the right sequence of buttons and switches to push? Might it be helpful if people were given a bit more background information about how a system operated, and what the buttons and switches were for? Would this extra information about the Taser bank help them make better sense of their task?

In a series of experiments, participants were trained either by practicing the steps of the procedure through rote repetition or by getting instructed in various versions of how the system supposedly worked.

The researchers found that not all types of information about how the system worked were equally useful to people. The information was only useful when it supported the user in making inferences about the *exact* and *specific* control actions they would need to take. People learned best in context, which supports the notion of trial and error as the best way to learn a system. If a user is taught about a device but fails to learn it correctly, misinterprets the model, or draws incorrect inferences on the necessary control actions, performance can actually be impaired. We need robots that are designed to guide people toward the information they need in order to decide how to respond in a given situation. The robot may not know *what* action needs to be taken, but still know that *some* sort of action must be taken, or it may know that an error has occurred, and direct the supervisor's attention to a particular part of the display, or walk the user through a certain sequence of actions to resolve the situation, like a checklist.

In the next chapters we will build our understanding of this approach. Core to effective teamwork is understanding your teammates' strengths and weaknesses—but not their general strengths and weaknesses, such as that they excel at basketball, but are not very good at cooking. We need to understand our teammates' specific strengths and weaknesses as they relate to working with us on our specific tasks at hand. And it turns out that we know a lot about humans' strengths and weaknesses when they are working with automation, and that knowledge will help us design new partnerships with working robots.

• CHAPTER THREE •

When Robots Are Too Good

IT WAS A SUMMER EVENING IN 2009 AND AIR FRANCE FLIGHT 447 prepared for takeoff to Paris from Brazil with 228 passengers.[1] The passengers boarded, settled into their seats, and waited. After they were airborne and the pilot announced they'd reached a smooth cruising altitude, passengers took out their books and laptops, or else reclined their seats to try to get some sleep. What they didn't realize was that the pilots were also relaxing. By now the autopilot was engaged, and the pilots were no longer needed. The captain decided this was the perfect time to take his rest period. He briefed the two copilots and retreated to the cabin.

Minutes later, ice crystals formed on the pitot tube, one of the airplane's sensors. Pitot tubes determine airspeed by measuring changes in pressure, and icing can block the flow, invalidating airspeed measurements. Because of this sensor failure, the autopilot automatically disengaged, and the plane entered a new flight control mode called alternate law 2b. Airbus flight computers can operate the aircraft in three different modes, or laws. Normal law is when all of the sensors and systems are functioning properly, and the control systems keep the plane within its specified flight parameters, preventing any brisk or large movements that could cause the aircraft to become unstable. When data is missing as a result of a sensor malfunction or other failure, the flight computer goes into either alternate or direct law. Alternate law is engaged when some of the protections are still functioning. Direct law is when there aren't any protections engaged and the pilot's inputs directly translate into movements of the aircraft's control surfaces. Alternate law 2b, the one that Flight 447 from Brazil entered on that day, is the most degraded of the alternate laws. It only goes into operation when something is so wrong that the pilot has to act without very much assistance from the automated systems. Of all the alternate-law modes, this one has the fewest protections against failures, including stalls.

But going into that mode wasn't the problem. The problem was that the pilots were back in control and didn't realize it. When the mode change occurred, the small green lines on their display next to the artificial horizon changed to amber crosses, to indicate the switch to alternate law. But the pilots missed this subtle change. Only once the plane started obviously malfunctioning did they notice something had changed, and in the heat of the moment, the pilots became surprised and confused. They lacked a full understanding of what had happened with the compromised sensor, and they had trouble compensating for the change in the aircraft's behavior. They tried to take over control, but they couldn't figure out how the autonomous flight controls were operating.

Within a few more minutes, the captain, called back in by one of the copilots, returned to the cabin. The aircraft was in distress, engine at full throttle, but rolling to the right and rapidly descending. The copilot who was at the helm was pulling back on the stick to try to get the aircraft to climb, but he didn't realize he was causing the aircraft to stall: he had climbed so steeply that air was no longer flowing over the wings properly. As the aircraft quickly descended, a different copilot was saying, "Climb . . . climb . . . climb . . . climb!" The one at the helm responded, "But I've been at maximum nose-up for a while!" Just then the captain realized what was happening and shouted, "No, no, no, don't climb!" All the pilot needed to do was go into a dive to regain airspeed and the wings would again become wings. But it was too late— the aircraft was too low to recover from a stall. Warnings started to go off as the ground was detected and the pilot said, "We're going to crash! This can't be true. But what's happening?" He was truly confused. The plane plunged into the ocean, killing everyone on board.

Decades of research in human-machine collaboration for aerospace applications shows that increased automation can undermine a human operator's ability to identify and recover from errors. Indeed, increased automation has led to many plane crashes.[2] Research consistently shows that when a system fails (and it doesn't matter how sophisticated the automation is—it will *always* fail eventually in some way), the people involved are worse off than if they had never used automation at all. The problems worsen when humans and automated systems must collaborate, especially when new systems are introduced.[3]

AREN'T MOST ACCIDENTS DUE TO HUMAN ERROR?

When catastrophic accidents occur in highly automated systems, the user is usually placed at the center of the blame. "Human error" is declared as the cause of the accident and deficiencies in the design are minimized; surely, the pilots of Flight 447 should have understood that the autopilot had disengaged, and been able to take the appropriate action quickly. After all, the light had changed from green to amber! Human error is blamed for many incidents not only in aviation, as shown in figure 8, but also many other settings. By some estimates, up to 90 percent of motor vehicle crashes are caused at least in part by human

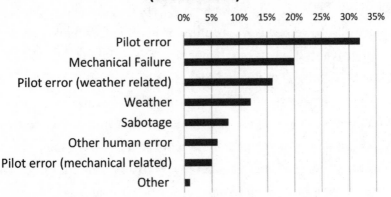

FIGURE 8: Pilot error is typically cited as the primary cause of aviation accidents. *Source*: "Statistics: Causes of Fatal Accidents by Decade," Plane Crash Info, http://planecrashinfo.com/cause.htm.

error.[4] In most self-driving cars to date, the human test driver has been found at fault for not reacting quickly enough. But is the person really at fault, when they are encountering a complex, unexpected situation that the developers of the system didn't necessarily foresee?

"Human error" offers a convenient explanation that protects the manufacturers of automated systems. But when you dig below the surface of the accident investigation results, a more complicated picture emerges. Most often, the real cause is a breakdown in the human-automation interaction. Sometimes, such errors simply reflect growing pains: when a new system is introduced that involves human-automation interaction, the humans have to learn new rules and gain experience with their new roles. But just as often, a breakdown can reflect the fact that the system hasn't been designed well. In the AF447 flight, the accident investigation group found that the pilots' standard, long-term use of the automated flight controls undermined their ability to quickly understand the situation when the automation suddenly and unexpectedly shut off. They had become too comfortable letting the airplane take care of things, and their skills for dealing with failures and sudden changes in the controls had degraded. The pilots had mentally checked out because they were relying on the flight deck automation, and they were not able to reorient in time to deal with the failure.[5]

One may then well ask whether this loss of "cognitive control" was really the pilots' fault. Could it have been in some sense a consequence of using a robot that was "too good"? What should we expect to happen when a machine works so well that the operator forgets how to intervene when things go wrong?

This possibility presents us with a conundrum. Obviously, we want to design robots that will work well, and the whole point of having them is so they can take over tasks from human operators. But this means that by design, the human operator will not spend much time operating the system in question or interacting with the automation. The operator will only step in when called upon to deal with a rare event failure. Ironically, the operator is needed only in high-stress situations, and high-stress situations, when there is little or no room for human error, are when people are most likely to make mistakes.[6] This problem is inherent to the process of automation. An early cautionary tale comes from our experiences in World War II. When designers improved the cockpit of the Spitfire plane, all seemed well and good during training. But under stressful conditions in dogfights, pilots tended to accidentally eject themselves from the cockpit. The root of the problem was that the designers had switched the trigger and ejector controls, and under stress, pilots reverted to their older, instinctively trained responses.

It is human nature that, in stressful situations, our perception of a situation narrows, our ability to act deliberately is compromised, and our actions deviate from what we would normally do. At the same time, the machine you're working with is also trying to respond to an event it's not used to dealing with. It's like when you get into a rental car for the first time and you're running late. You've started a car, adjusted the seats, and driven many times. But never in this car. And because the controls are slightly different, you struggle to get oriented and out of the parking garage, often onto streets that are unfamiliar to you, which only increases your stress level. With autonomous systems, the problem is very similar. The system is the same overall, but the details have changed. This puts the operator in the worst possible position for making quick decisions and taking precise actions.

The problem is compounded by the fact that automation frequently conceals the behavior of the system. This is done on purpose, to make interacting with complex machines simpler and more manageable. But as we've seen with AF447, it leaves the user struggling to figure out what the machine is trying to do, whether that is the right thing for it to do in the situation, and what commands to give it to bring the system into a safe state. In the rental car example, you can adjust the seat and the mirror, find the switch for the headlights and windshield wipers, and be on your way. But with autonomous controls, there is minimal information on the display and there may not be any physical representation for how any one mode works. You have to be trained or learn it through use. While users' understanding of the automation modes and behavior improves the more they use it, the modes a system enters during a crisis are by nature only rarely experienced and therefore remain unfamiliar until the moment they are needed.

This is a problem for more than just pilots of complex aircraft. Now that most of us are using GPS, for example, we may be losing our navigational abilities. We listen to the calm voice of the GPS system, which patiently gets us back on track when we make mistakes, and we get to where we're going—but many of us no longer have a mental picture of our route from here to there. This is okay until we enter a situation that requires that deeper understanding, like a road closure. We quickly fumble with GPS to reroute or find other roads that will get us back on our predetermined route.[7]

Through studies prompted by airline accidents, we have learned that it's only human nature to become complacent with automation that functions so well that we come to rely on it—but we can quickly come to rely on it too much at some times while not trusting it enough at other times.[8] In laboratory experiments, many people continued to use an automated system even after it failed because they had developed a blind dependence on it.[9] Even when we're aware of the risks, we may still lose focus and struggle to pick up on the subtle cues that the system may be approaching a failure state. Since failures in automation occur so infrequently, people rarely detect them in time. This phenomenon, known as the *vigilance decrement*, has been seen across many rare event detection tasks because people lose their concentration in the task as time passes without an event.[10] Vigilance decrement persists no matter how familiar a user is with a system.

The problem of inappropriate reliance is even more severe when we interact with a more peerlike or anthropomorphic robot system. In a recent study from Georgia Tech, researchers tested a robot meant to guide people safely out of a building that was on fire.[11] People seemed to follow the robot blindly, even when it was very clearly malfunctioning or leading them into dead ends inside of closets. The problem is exacerbated by our historical fixation with making robots look and act more like humans. Through many studies, we have seen that the more anthropomorphic the system is, the more likely a person is to inappropriately rely on its suggestions, advice, or actions. In other words, people trust robots too much for their own good. We need some way to approach the design of these new systems that preempts these problems, rather than waiting to learn from our mistakes as we did in aviation.

Pilots are required to consistently perform at the boundaries of their cognitive capability. They continuously monitor the complex aircraft and must be ready to act reliably and quickly when something goes wrong. If they blink, or their attention is divided at the wrong moment, the results can be catastrophic. And therefore it makes sense when designing aircraft to carefully study the strengths and limitations of human cognition and information processes and account for them in the design.

But you are likely not an elite airline pilot supervising an advanced autopilot system. So what does this have to do with you?

If you are a parent or have ever babysat a toddler for an extended period of time, you are essentially doing the job of a modern-day airline pilot. And the stakes are just as high—it comes down to the safety and well-being of your child. If your two-year-old is like the ones we know, they must be watched constantly. The moment you turn your head away, they will make a beeline straight to the bookshelf and start climbing. Do you ever wonder why it's so mentally exhausting to watch a toddler? It's a long-term monitoring task requiring constant vigilance. And people just can't sustain such intense monitoring tasks well for very long.

This is why, whenever possible, we try to minimize long-term monitoring in safety-critical jobs. In operating rooms, nurse surgical assistants have much shorter shifts than the surgeons themselves, for example, taking breaks during surgeries longer than two hours or so though the surgeon may be at work for twelve hours or more. That is because the surgical assistants are primarily monitoring the surgeons—providing the right surgical instruments at the right time, mentally tracking the instruments and sponges to make sure nothing is left in the patient. A person just can't do that type of monitoring task very well for very long. Transportation Security Administration (TSA) agents inspecting our luggage at the airport rotate frequently for the same reason. Air traffic controllers take a long break after only thirty minutes or an hour of monitoring and directing planes as they cross our skies.

Unfortunately, parenthood does not offer the same relief! Accidents are a leading cause of death and disability during childhood. Each year, twenty-three million kids under the age of fifteen end up in the ER— from drowning, poisoning, choking on small toys, dog bites, playground injuries, injuries from exposure to household safety hazards, fall-related injuries, and the like.[12] And it's not due to supervisory neglect—most accidents happen when a caregiver is supposedly watching the child.[13]

It's just that we are not machines, and our attention will sometimes lapse. Indeed, one of the reasons we value robots is that their attention spans are essentially infinite! Parents have to occasionally accomplish other activities of daily living—walking away for a moment to answer a call, cook dinner, or empty the dishwasher. We must constantly think and decide—will my child keep doing what he's currently doing if I step away? How long can I wait before popping back to make sure he is still safe? The longer we are gone, the more our *situational awareness* degrades.

Situational awareness (SA) is a theoretical construct that helps us understand how people process information, understand complex systems and situations, and develop the ability to make decisions.[14] We have been studying how our interactions with automated systems affect our situational awareness for decades. This research helps us better understand our typical processing strengths and limitations as humans. Moreover, the findings on situational awareness provide a foundation for improving the design of systems involving human-automation interaction. A separate and complementary line of research looks into how people make quick decisions with limited or uncertain information. People tend to apply *heuristics*, or shortcuts, in decision-making that lead to predictable biases in judgment.[15] Poor human decision-making due to these kinds of biases has been documented across many applications, including economics, marketing, fantasy sports, firefighting, and, of course, aviation.[16] The typical weaknesses in human situational awareness, along with heuristics and biases in decision-making, represent some of the biggest, most common holes in the average person's Swiss cheese model. Understanding these problems will be key to designing good robotic teammates.

Studies say that when parents are supervising their children, they are constantly performing a complex mental calculation about what interventions seem appropriate.[17] Their decisions will depend on the nature of the child, the current activity, the child's age or developmental status, the child's gender (there is a sense that boys are less predictable than girls, and are left alone for shorter periods of time), and many other factors. And different caregivers have different functional parenting competencies and styles of supervising. Some parents, dubbed "helicopter parents," hover over their child's every move and prefer to prevent children from putting themselves into a compromised situation. Other parents practice permissive parenting and believe that their children learn more by figuring their way out of compromised situations independently. The parenting style and level of competency ultimately affect a child's safety and well-being. Simply put, some people are naturally better at monitoring children than others, and some try harder at it than others. The same is true of people in the professional world. Two physicians may very much enjoy the operating room, but the one who is very good at monitoring tasks may naturally gravitate toward becoming an anesthesiologist, and the one who is less suited to monitoring tasks may naturally gravitate to the more active role of surgeon. Again, parenthood gives us no such choice.

Our workload and fatigue levels also impact our ability to effectively supervise our children. Single parents have many balls to juggle, and they have to do it with less help.[18] The data indicates that a child's risk of injury increases substantially when the child lives with a single caregiver, or in a home with multiple siblings. And that finding only makes sense, because these situations decrease a caregiver's cognitive capacity to attend closely to a child's activities. The same is true for people interacting with highly automated systems: the more overloaded or tired people are, the more likely it is that they will miss key information or make a mistake.

But there is some good news and bad news for parents. The good news is that a number of companies are developing and fielding home robots to help parents supervise children.[19] Currently, they are used as remote monitoring devices. The robot can follow the children around the house, its camera pointing toward the action, while the parent cooks dinner in the kitchen, keeping one eye on the remote video feed and an ear on the audio to listen to the children play. The vision is that these systems will soon become more intelligent. They will track a child's body motions and head gaze, and may be able to alert the parent if the child is starting to climb on the furniture, or is gazing inquisitively at an electrical outlet. In other words, you can share the monitoring task of parenting with the robot, and it can help you decide how long you can stay away and when to check back.

Problem solved? Not so fast.

Did cockpit automation get rid of pilot errors? Some, yes, but just as aircraft automation introduced new complexities and created new problems, so, too, will the introduction of babysitting robots. Your task as a parent will shift from monitoring your child to monitoring the ability of the robot to monitor your child. You've spent years understanding your child, building your mental model of how he or she will behave, and learning how long you can look away. Now you also have to understand how the robot is functioning, what it can do well, what it might miss. The robot may have multiple "modes" you need to keep track of—a mode for safety in the house versus on the playground, for instance. Now you have to worry and check whether the robot is in the right mode. And, perhaps most importantly, if you rely on the robot babysitter too much, you might not notice when something has gone wrong, and become unable to intervene in time. The system will work only if parents truly understand its strengths and weaknesses, so that they rely on it only at the right times and use their own judgment at other times.

Getting this right will not be easy, but again, designing such systems will not require us to reinvent the wheel. Human-robot collaboration can find guidance in similar challenges that designers of industrial systems have already faced. We can draw from the science of human cognition, which has helped to make air travel one of the safest forms of transportation. Let's look at the theory of human situational awareness; at the ways in which workload, fatigue, and natural human biases affect human performance; and at how technology complicates our ability to make good decisions.

We know that accidents spike when we introduce new technology into aircraft, and we have refined our theories of human decision-making when using complex technology the same way we have refined our robot technologies. Let's make sure that here, again, the lessons and insights from one domain are transferred to the others. Can we prevent accidents from happening as we introduce home robots that are more capable, by learning from our past hard-earned lessons? We cannot afford for injuries to children to spike beyond the twenty-three million per year they already suffer, just because we were not able to transfer what we've learned from industrial applications to consumer products.

SITUATIONAL AWARENESS RESEARCHERS STUDY HOW WE MAINTAIN MENTAL representations of the systems in the world around us. They are interested in our interaction with all types of systems, but their work is especially useful in designing complex systems. It enables us to predict how human actions are likely to affect a given situation, providing the basis for effective decision-making. Developing and maintaining situational awareness involves three stages: perception, comprehension, and projection. Perception involves processing the sensory information of elements in the world around you. Comprehension involves synthesizing these new inputs and comparing them with your goals, expectations, and current model. And projection is the use of this information to predict future states and outcomes.[20]

When an autonomous machine cedes control of a task to its human partners—for whatever reason—it is essential for those human partners to be at attention and that they maintain adequate situational awareness.

Let's return to the example of driving a car (just a regular car this time, not a self-driving one). When you drive, you are receiving all kinds of visual inputs simultaneously. The dashboard of the car is telling you how fast you are going; you have to watch the road and be aware of what's ahead of you, or on intersecting side streets; and you take note of landmarks, the weather, and so on. And then there are aural inputs: other people in the car chatting, perhaps, along with the hum of your engine, honking from other cars, your GPS giving you instructions, or your Spotify playlist. And tactile input, such as the vibration of your car, the pressure of your foot on the gas pedal, and the changes in acceleration as you speed up or slow down. You subconsciously fuse these inputs together to understand when you need to slow down, when to turn, or when you're near your destination. And you instinctively know how hard to push the brake pedal to slow down in time for the traffic signal without having to screech to a halt, how to adjust your driving behavior when it's snowing, when to watch out for pedestrians, and when to pause in a conversation with your passengers so you can focus more on the road.

Imagine if you lost your situational awareness while driving. Let's say your dashboard instruments went down or your rearview mirror broke off. You would struggle to gain the same level of understanding you are able to have with those aids. You would either compensate by getting the same information in other ways, or you would allow your situational awareness to degrade. For example, to work around a broken rearview mirror, you would use extra caution, perhaps looking over your shoulders more often. And that would suffice, most of the time. But on rare occasions, the struggle to maintain adequate situational awareness could cause you to make mistakes, and potentially lead to an accident. This is a simple example, but as robots enter our everyday lives, we will offload increasingly more tasks to them, and if this handoff is not designed well, the transfer of work to robots may make us worse off than if we'd just kept those tasks for ourselves. The loss of overall situational awareness could leave us unempowered to catch errors or intervene when needed. For example, if a robot in the hospital is responsible for delivering medicine to different rooms, and then it becomes stuck on an obstacle in the hall, who will recognize that it's stuck, free it, and make sure it gets back on track? Or, even more urgently, who will make sure the waiting sick patient will get the medicine through another means?

On board the Air France flight that began this chapter, the pilots were unable to develop adequate situational awareness as the aircraft descended into the ocean. And we have seen this kind of scenario play out again and again. In another incident, China Airlines Flight 006, pilots struggled to understand the orientation of their aircraft, and as a result the aircraft plunged thirty thousand feet within minutes.[21] One of the engines had failed, and as the flight engineer worked through a checklist to understand and remedy the problem, the automation attempted to compensate for the asymmetry in thrust from its engines and turned the control wheel as far to the left as possible. The pilot needed to understand that this was happening before shutting off the autopilot, so that the crew could prepare to compensate for the problem manually. However, the captain failed to apply the correct controls before or after disconnecting the autopilot. Because the plane was passing through clouds, the pilot needed to rely on the artificial horizon indicator on the instrument panel, which displays the aircraft's orientation relative to Earth's horizon. The instrument displayed the excessive bank and pitch—that the plane was essentially flying down and on its side. Because this was so unusual, the pilot assumed that the indicator was faulty, and he became spatially disoriented. It was only after the aircraft broke through the clouds that the pilot was able to reorient himself and realized that the attitude indicator was correct. Once he regained his situational awareness, he was able to recover the aircraft, restart the engine, and land safely.

The loss of situational awareness while operating highly autonomous systems has accounted for hundreds of deaths in commercial and general aviation, and as a result, the government and the aviation companies have spent untold millions of dollars studying why this happens and trying to develop solutions. The goal is often to figure out how to prevent the loss of situational awareness.

Weaknesses in someone's awareness can stem from problems in their perception of data and specific elements in the environment. Alternately, someone might perceive all of the relevant information, but fail to comprehend its meaning. And finally, someone might perceive and comprehend correctly, but then fail to use that information to accurately project future information or events. There are many possible underlying causes for failures at each of the three levels that system designers must consider, and some of the most common are listed in table 1.

TABLE 1
TAXONOMY OF SITUATION AWARENESS ERRORS

LEVEL 1: FAILURE TO CORRECTLY PERCEIVE THE SITUATION
Data not available
Hard to discriminate or detect data
Failure to monitor or observe data
Misperception of data
Memory loss

LEVEL 2: IMPROPER INTEGRATION OR COMPREHENSION OF THE SITUATION
Lack of or poor mental model
Use of incorrect mental model
Overreliance on default values

LEVEL 3: INCORRECT PROJECTION OF FUTURE ACTIONS OF THE SYSTEM
Lack of or poor mental model
Overprojection of current trends

The likelihood of making mistakes depends in part on how much the user is paying attention, increasing with either too much or too little engagement.[22] This means that asking someone to pay either

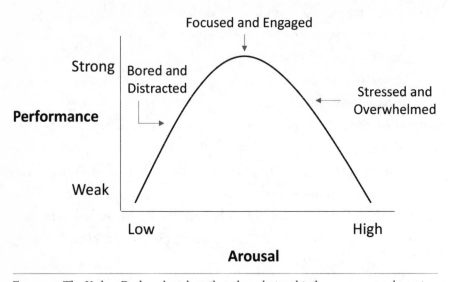

Figure 9: The Yerkes-Dodson law describes the relationship between arousal or stimulation levels and a person's performance on a task. *Source*: Robert M. Yerkes and John D. Dodson, "The Relation of Strength of Stimulus to Rapidity of Habit-Formation," *Journal of Comparative Neurology and Psychology* 18, no. 5 (1908): 459–482.

high amounts or low amounts of attention to automation can be counterproductive. There is a nonlinear relationship between "arousal," or the level at which someone allocates physical or cognitive resources to a task, and the ability to successfully perform a task. This relationship is called the Yerkes-Dodson law, named after the psychologists who first described the phenomenon. As shown in figure 9, when users do not have enough stimulation, they become bored and are more easily distracted from their primary task. But when they are overstimulated, they eventually become stressed and overwhelmed. During these moments, cognitive processes actually change. According to the spotlight theory of attention, we typically will have a tremendous focus on a narrow part of our task, which causes us to completely ignore or overlook visual stimuli that are outside of our focused attention zone. Automation is typically designed to reduce the task load on users, which by design removes their participation in routine activities and only engages them in high workload activities, such as recovering from failures. So users of automated systems are in the worst parts of the performance curve. Optimal performance is about finding the right balance of tasks between users and machines.

One obvious solution might be to simply make robots less automated, thereby requiring users to pay more constant attention to them. This choice would ensure that users remained engaged and ready to take over if and when the system failed. But one must be careful to consider the users' total workload—the work involved in all the different tasks they perform—not just the workload involved in interacting with a robot. If you turn back the dial on automation, and the workload of the users ends up being too high, their decision-making abilities decrease and their performance degrades. As a result, they are less able to take over when an emergency situation demands it of them. This is what happened in the cockpit of the Air France flight when the pilots attempted to understand the situation and the state of the aircraft.

We also have to consider the cumulative effects of fatigue. Fatigue has both physical causes (e.g., sleep deprivation, hunger) and mental causes (e.g., time on task), and studies have shown that decision-making can change depending on fatigue level.[23] When judges issue rulings, for example, they are performing a mentally taxing task. They have to take in and comprehend a lot of information and project possible future events to make their decisions. And fatigue affects the outcome—there is evidence that judges tend to be substantially more lenient in their rulings right after taking snack or lunch breaks.

It also matters how detailed a task is. There is a difference between experiencing a high workload because you are performing a really challenging task and experiencing a high workload because you are engaged in multiple concurrent activities. People consistently have a slower response when they have to switch between tasks. One laboratory study asked participants to switch between solving math problems and classifying geometric shapes. Participants lost time when they switched between these two tasks, and the lost time increased as the tasks became more complex.[24] We switch tasks all the time—for example, we reach down to change the song on the radio while driving and then go back to monitoring the road. We check a text message and then look up to make sure our children are safe as they climb up a sliding board. Hundreds of milliseconds are lost as we switch between these tasks because our cognitive processes must readjust to perform each of them differently. It's as if our brains must stop and quickly load a new program. While the switching costs are often small, usually less than a second, in safety-critical applications those delays can make a person incapable of responding in the time necessary to avoid an accident. Mental context switches resulting from task switching are also known to undermine our situational awareness, compromising our ability to correctly respond to a hazardous situation.[25]

Many of the robots in our lives today are limited to a single task. But more advanced automated systems are typically designed with multiple modes to handle different kinds of unique situations. More and more, operators are going to have to understand the different kinds of behavior for the automation in each of several different modes, and this switching is going to be another pain point.

Say that your car, like all other new models, has an antilock brake system (ABS), for example. In good weather, you press on the brake pedal and your car applies the brake to slow down. However, in bad weather, your car operates in a different mode. Depending on the road conditions, it may not directly apply the brake as you press the pedal. To avoid losing traction with the surface, it will apply and release the brake in pulses. In both cases, you pressed on the brake pedal in just the same way, but the car's two responses felt different. If you didn't know in advance that your car had an antilock system, the second result could be very surprising. As an instinctive response, you might worry that your brakes were not working properly and take an inappropriate action, such as swerving the car. The early antilock braking systems alerted the driver about ABS activation by vibration through the brake pedal. Drivers were confused and alarmed by this unexpected signal, and many of them responded by taking their foot off of the brake, which was not the correct response.[26] This simple form of mode confusion became the cause of many road accidents.

Aircraft have even more modes than cars today, and operators' confusion about automation modes is a substantial contributor to airline accidents.[27] On the AF447 flight, the autopilot disconnected because the airspeed measurements were faulty, and the aircraft transitioned from normal law to alternate law, which does not provide the same stall protection that pilots are accustomed to under normal operation. The pilot struggled to understand how his actions were being translated into inputs to the aircraft and continued to pull back on the stick, essentially causing the plane to stall without realizing it. The stall warning sounded for fifty-four seconds, but the pilots did not comment on the warning and did not seem to realize that the aircraft had stalled, because they were distracted trying to figure out what was wrong. The warning sound may not have been clear, or their spotlight attention may have been focused on the wrong set of alerts, causing them to miss the most urgent, fundamental issue. Clearly, they did not have accurate situational awareness and therefore were not equipped to make the right decisions in that moment.

While aircraft automation may still arguably be more complex than that of cars, soon this may not be the case. The same safety issues due to mode confusion are beginning to crop up in our assistive-driving cars. A 2017 study of the Tesla Model S 70 documented eleven instances of mode confusion for a single driver over a six-month span of driving.[28]

And clumsy design of automation can make a bad situation worse. If a system does not display or clearly communicate its mode to the operator, it can lead to misinterpretation of information and inappropriate actions. The same problem can arise if the system communicates its mode to the operator, but in a manner that confuses the operator. For example, in 2017, on board the USS *John S. McCain*, a US Navy destroyer, the commanding officer decided to redistribute control and ordered throttle to be transitioned to another watch stander's station.[29] But the helmsman accidentally shifted *all* controls over to the new watch stander's station. When that happened, the rudder was automatically reset to its default position (centerline of ship) without any warning or notification to the operator. The mode change took the vessel off course and put it on a collision course with a commercial tanker. Everyone on board thought there had been a loss of steering, and it took several minutes for the crew to figure out what was going on. But it was too late, and the USS *John S. McCain* collided with the tanker, killing ten sailors.

Mitigating mode confusion requires careful work on the part of the designer. The user will need to be able to understand the robot's behavior sufficiently in any given situation. The automation should provide clear cues about the system's modes, states, and actions. It should not provide feedback that could be misinterpreted during key moments—that is, at moments when the operator's correct input is imperative. And in case the user takes the wrong action, the automation must also be smart enough to identify and respond to mistakes in operational decisions.[30]

The operation of safety-critical systems often requires rapid decision-making by those using the system. If a person is deprived of resources—that is, they do not have enough time, information, or cognitive capacity to make logical, well-analyzed decisions—they'll take shortcuts. Typically, they apply heuristic principles, which reduce the complexity of the problem, and make quick judgments based on experiences they have had in the past.[31] This is often a powerful way to make decisions when your situational awareness is impaired. However, like all models, heuristics can contain significant biases.

A common heuristic is the *availability heuristic*, which is when people judge whether an event is likely based on what they can most easily recall or imagine. For example, during the China Airlines incident, it was easier for the pilot to imagine that the attitude indicator was faulty than to think through why it might be presenting such extreme vehicle orientation data. The heuristic enabled the pilot to come to a quick judgment, but unfortunately it was inaccurate.

Nearly a hundred other kinds of biases have been identified and experimentally validated. But these human decision-making vulnerabilities are rarely considered when robots are designed, because engineers rarely consider human psychology during the design process. As a result, new systems can easily trigger or amplify these biases rather than compensating for them.

Another obvious way to counter these concerns about making robots that are "too good" is to simply train people better. According to this thinking, if the operators could be better armed with information about the robots they used, disasters like the collision of the USS *John S. McCain* would be avoided. But if it has not yet been made clear, we would like to state unequivocally that people have fundamental limitations and common biases that will occur regardless of how much training they are given. Even more to the point, as new automated systems continue to enter our everyday lives, we need to design them under the assumption that the average consumer will not have much training on them, if any. We will all be operating and interacting with working robots that we do not fully understand.

WON'T ROBOTS KEEP GETTING SMARTER?

We can't change human nature, but the hope remains that better technology can solve the problems that technology creates. Those who take this view think that if only we can make our automation more intelligent, or make our robots smart enough to act more like human partners, then surely these problems will disappear. Of course, this would require making a truly humanlike robot, one that can learn as well as humans and read all our gestures, facial expressions, and other implicit cues, and that can communicate with us as easily and naturally as we do with other people.

The bad news is that even if you made the most humanlike, intelligent robot partner possible, the problems would still exist. Think about pilots in the cockpit. You can't get more humanlike than real humans. And human pilots are elite teams: they have thousands or tens of thousands of hours of experience flying. Yet 73 percent of airline accidents happen on the first day that a new pilot-copilot crew is working together, and 44 percent in the first flight of that first day.[32] The reason is that being human does not automatically make you easy to work with. People—even highly trained pilots—need some experience and practice working together with a new teammate. As they work together, they build mental models of each other and learn to communicate effectively. We can't rely on the average consumer to have extensive training in operating the next generation of robots, and we can't expect them to know how to interact with them. We need new approaches to designing these systems so that they allow a novice user to quickly build a mental model of their operation and behavior.

But what if we built robots with superhuman capabilities? Maybe superintelligent robots could analyze scenarios more clearly than humanlike ones and be granted the authority to decide who should do what when. Couldn't this approach make human-robot teams more capable than human teams? To be fair, robots have recently become capable of working with people in this way. In fact, our lab at MIT conducted a series of studies looking at the potential benefits of a "robot boss" versus a human boss.[33] Two people and one robot worked together to gather materials and assemble a structure out of Lego building blocks. The Lego assembly task was designed to require collaboration among the humans and the robot, with trade-offs in the team's effectiveness depending on who retrieved which parts, in what order, and who did the building. The robot was given information about how well the human members performed various aspects of the task, and it then employed advanced planning techniques to optimize the work allocation and direct the human teammates. This case was compared with two other cases, one where a person fully directed the work, and another where a person and a robot shared authority over tasking the three-person team.

The results were surprising. First, at least in these studies, it seemed that people actually *preferred* the robot to make all the decisions![34] This may be because the robot was clearly better at the task, producing better work schedules overall—and conversely, decreasing the robot's authority over work decisions reduced the team's efficiency, at the same time decreasing the desire of the humans to work with the robot. This preference for increased robot intelligence and decision-making authority has been identified in other settings, too.[35]

The benefits nevertheless came at a steep, arguably unacceptable cost. We found that as robots made more decisions, the human team members lost awareness of their teammates' actions. People were less aware of what their human and machine partners were doing and what they were going to do next. We know from the long history of accidents in aviation that this lack of situational awareness undermines a person's ability to take over when a robot fails.

Furthermore, we know that when a human partner is working with an intelligent robot, the human actually needs to know *much more* about the system, not less, than they needed to know about traditional automation. The reason is that the robot is an independent, at least semi-intelligent, entity with different objectives and plans than the user. It's not enough to know what mode the system is in: the person also needs to know what the robot is trying to do and why, if he or she is to effectively team up with that intelligent agent.[36] Even with the most advanced robot technology, we still see that increased robot autonomy increases the risk and costs resulting from reduced situational awareness.

Finally, as you can imagine, these difficulties get worse if the automation changes its behavior over time, helping the person less or more depending on the circumstance. The Internet of Things makes it possible for companies to upgrade the software anytime your device is connected. This makes it easier to address flaws in products and improve them. But it can also create confusion for the user as the device's behavior changes or new features appear. Yes, it's often possible to go online, read about the update, and find out what it included—but we don't actually do that. This issue has become a popular topic of discussion on forums for operators of commercial drones, who are concerned that updates increase the risk for crashes. Automatic updates of Tesla cars have also caused confusion and even dangerous situations for drivers.[37] Companies strive to improve safety with each upgrade, but the clumsy introduction of new versions may cause new and unexpected human-automation interaction breakdowns.

We see a similar problem with new machine-learning technologies that make it possible for robots to change their responses to the environment or to the user. If a robot changes its behavior over time, it can turn out to be less than helpful, as users have to spend mental energy trying to understand the changing behavior.[38]

OLD PROBLEMS AND NEW CHALLENGES

We've learned hard lessons from our decades of experience with automation in aviation and other safety-critical fields. The cost of these lessons is measured in lives. Early evidence indicates we will see many of the same challenges as we introduce the new breed of working robots into our cities, onto our roads, and into our homes. Fortunately, we have a starting point to preempt these problems, rather than having to learn from our mistakes as we did in aviation. We have a foundation of theories, empirical data, and design principles that can inform the design of our intelligent machine teammates and ensure that these systems shore up, rather than amplify, our human weaknesses.

This body of work tells us that people have fundamental limitations when working with complex automation in time-critical situations, and that robots must be intentionally designed to overcome these limitations. We must also effectively prepare users to intervene in areas where these systems have weaknesses. In tasks where user intervention may be needed, the work of robots must be held back, to allow users to remain consistently engaged and maintain situational awareness—or else they should be kept out of the decision loop altogether. There is rarely an effective approach in between. Moreover, situational awareness must include a robust understanding of the modes of the automation, which means that the automation must be designed to make these modes clear. Too much complexity with the modes makes them too confusing. The number of modes, for example, should be minimized, so that there are only as many as necessary. Transitions between modes must be simple and clear, to support the user in having an intuitive understanding of when mode changes are occurring.

These next-generation systems bring with them new questions and challenges that are unlike any we have previously faced. We have no prior experience with intelligent robotic systems that operate in such close proximity, on the same tasks that we're engaged in ourselves, interacting with the complex dynamics of everyday human experience. Studies on human-automation interaction to date have been relegated to small numbers of operators and one or two automated systems. They have been conducted in settings that have been carefully engineered by humans, such as cockpits or control rooms. And so far, these systems work according to well-codified rules, whereas our new systems are continually updating and learning. And what the system is learning, or how the behavior is changing, is seemingly impossible to communicate to people, because they will receive little or no formal training on these systems.

There is even another curveball coming toward us. Up to this point, our studies and insights regarding human-automation interactions assume there is a human operator or supervisor managing the technology. Designers have this in mind as the basis for potential interactions between human and robot collaborators. But the intelligent robots released on our streets and sidewalks, and into our workplaces and homes, will necessarily interact with many people who are not "designated operators." What about them?

CHAPTER FOUR

The Three-Body Problem

EGINNING IN 2016, A HANDFUL OF TECH COMPANIES started rolling out new sidewalk delivery robots on the streets of San Francisco, a very tech-friendly city with a high percentage of early adopters of new technology. Residents ordered food from local restaurants, and robots would deliver it.

In December 2017, city officials responded to a public outcry from residents.[1] Some residents had sent photos to their local officials of the robots hogging the sidewalks. Senior citizens and people with disabilities were particularly concerned, as they felt unsafe maneuvering around robots that didn't understand their needs. The city imposed strict rules for how robots could be used. They limited the number of delivery robots in the city to nine total, confined them to nonresidential areas where they would have minimal interaction with people, and required them to have human escorts.

We will see this same interference of robots in our daily lives in other cities as they become more pervasive on the ground and in the air. They will cross our paths on the sidewalk, merge in front of us on the highway, board elevators we are riding, and land at our doorstep. And although we will all likely use working robots at some point, most of the time when we interact with them, we will be merely bystanders. This is even true for most of the humans we meet on any given day. The crowds you navigate on your way to work, the people waiting with you in the drugstore checkout line—they are bystanders to your own plans, as you are a bystander to theirs—and certain norms dictate how we behave when we come into contact with each other. In this chapter, we'll explore how to plan for arguably the most common type of social interaction a working robot will need to be prepared for: communicating their intentions to us when they're not working with us or for us directly.

Psychologist Donald Norman argues that there are two basic principles for designing objects to be used by people: (1) provide the user with a good conceptual (or mental) model of how the product works, and (2) make things visible.[2] He offers scissors as a basic example, because there is a visible relationship between the design and the use. You could pick up a pair of scissors and know how to use them even if you'd never seen them before. The holes are rightly sized for two fingers and a thumb, and when you hold them and move your hand, the blades open and close, and you can easily see how to use them. You don't need any training or instructions. By contrast, typically, digital watches are poorly designed for people. It's not easy to see how the buttons on a watch relate to the watch's functionality. It takes a lot of trial and error, pressing the buttons and observing the response, before a user can figure out how to set the time and so on. Remote-control devices for televisions attempt to solve this problem by including lots of buttons with labels or symbols on them, such as up and down arrows, to indicate their functions. But they are still confusing to many users.

A system's designer has to create the system's conceptual model, and no doubt most designers want their systems to be easy to use. But they are not always successful. It isn't always recognizable, in the phys-

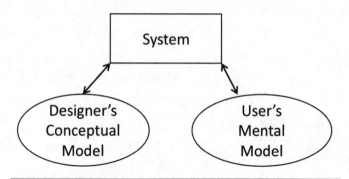

FIGURE 10: Diagram showing how a designer's conceptual model and the user's mental model of a system can be two separate things; both intend to represent the system but neither is an exact model. *Source*: Adapted from Donald Norman, *The Design of Everyday Things* (New York: Doubleday, 1988).

ical design of a system, how it is supposed to operate and what its constraints are. The design should inform the user's mental model of the system, but many designs only complicate the formation of an accurate model. Figure 10, adapted from Norman's work, depicts the designer's and user's mental models as separate constructs.

Modern robotic systems are the most complex systems people have ever made. They are systems that can learn from observation and through interaction with the environment and make decisions on their own. They have intricate interfaces between humans, the environment, and other technologies, and these complex interactions can result in robot behaviors that designers and users did not expect. The user of a robot, as suggested in the previous chapter, is perhaps more accurately understood as a *supervisor*, because there is an expectation that the robot will operate with a degree of autonomy that the user simply monitors, intervening only when necessary. The supervisor may physically interact with the system or may be hundreds of miles away, monitoring it remotely. Nonetheless, users of robotic systems are still, in some way, responsible for the robot meeting its goals, and they need a high-quality mental model of the robot system's operation in order to carry out that role.

Outside the controlled environments of labs or factories, there is another actor that interacts with robots, separate from the designer and user. That person is the bystander, someone who comes in contact with the robot but does not have responsibility for it or wield direct control over it. Bystanders have different goals than robots—they are operating near each other, but not in concert with one another. They may not even be aware of each other. Bystanders build their own mental models of robots based on their limited interactions with them. But this mental model is different from the designer's and the user's, because bystanders do not care about the robot's success in the same way. There is a concept in physics called the *three-body problem*. It has to do with how gravity affects large objects (like planets and suns) when there are more than two of them—the mathematics become much more complicated when you are trying to calculate the effects of gravity on three bodies than when you are trying to calculate the effects on only two. In robotics, we can adopt this term to refer to the problems that arise with three participants in the human-robot interaction. Instead of just a robot and a user/supervisor, we need to consider the robot, the user/supervisor, and the bystander. And here, too, things become a lot more complicated.

For example, security robots are increasingly patrolling apartment complexes, sporting events, and neighborhoods. They incorporate self-driving technology, laser scanners, thermal imaging, 360-degree video, and many other sensing capabilities. They drive around areas needing heightened security and offer a robust physical presence to deter theft, vandalism, or other security threats, capturing various kinds of data to increase the likelihood of identifying criminal activity. A human supervisor may work in collaboration with one or more robots from a distance, monitoring the collected data in real time and redeploying security staff based on his or her observations. These users—or supervisors—develop their mental model of the robot by reading manuals and undergoing training. But the residents of the neighborhoods and patrons of the shops and businesses where the robot is on duty will be surprised at first to see this unidentified robot patrolling their sidewalks. They must rely on casual, brief observations to figure out why it's there, and wonder whether it might run over their toes, harm their children or pets, or even misidentify them as criminals. In July 2016, a security robot patrolling the Stanford Shopping Center in Palo Alto, California, knocked down a sixteen-month-old toddler and then continued on its way.[3] According to the manufacturer, the little boy had left his parents' side and begun running toward the robot, probably interested in the strange-looking machine. The robot veered to the left to avoid running into the little boy, but the boy also changed directions, and the two collided. In this case, there were no serious injuries—but it could have been worse: the robot weighed three hundred pounds and stood five feet tall. This is the type of social dynamic that can be very challenging for robots to manage successfully. The parents may not have even noticed the robot until their child was already intrigued and heading for trouble, and the robot was unprepared for the child's unpredictable movements. Instead of changing direction, it would have been better for the robot to stop in this case. A human security officer would probably have paused to interact with the child, or would at least have known how to maneuver around a child. Shouldn't the security robot have been able to show the child the same attention or accommodation that any other social entity would have shown?

Because bystanders do not have as much direct contact with a robot as a trained user, they operate with a very limited, and perhaps inaccurate, mental model of what the system could conceivably do. Also, because their goals are different from the robot's, the robot may be merely a distraction, and potentially a nuisance, to them. In general, robots should minimize interference with bystanders as much as possible. And when interference does occur, the negotiation process between a robot and a bystander should be as quick and easy as possible, similar to how we maneuver past people in a crowd with simple phrases such as "Excuse me." We don't need a detailed understanding of what others in a crowd are doing or where they are going to maneuver around them. But we do consider high-level information, such as whether they are elderly or a child, are in a lively conversation and may not see us, or have something heavy or awkward in their hands, such as a hot coffee or two. When we take note of high-level information about others around us, we subconsciously factor it into our movements and decisions. If there is an elderly person crossing in front of us, we give them more room; if a parent pushing a stroller is trailing behind us as we approach a store, we hold the door open a little longer than we would otherwise, to help them enter comfortably. We accommodate others all the time without exchanging any detailed information with each other.

By the same token, bystanders don't need a detailed understanding of a robot's global plans, intentions, or reasoning. We can contrast the needs of the supervisor and the bystander in terms of the *active* and *passive* perception systems in the human brain, one driven top-down by goals and the other driven bottom-up by stimuli.[4] The supervisor, who is the robot's teammate, requires an *active model* of robot behavior, developed by a search for information that is focused by the supervisor's and robot's shared goals. Bystanders only need enough information to do their part to minimize interference and negotiate the brief interactions that occur. They are no more responsible for directly monitoring a robot than they are for monitoring a person crossing their path. That is, they only require a *passive model* of the autonomous system's behavior, which is derived from observing and processing information about the system as it behaves in pursuing its own goals.

For example, you may take a quick glance at someone coming toward you on the street and assess that person's body language to decide whether the person is likely to change direction as you pass each other. You do this without consciously thinking about it, and the assessment takes place in a fraction of a second. But if you noticed that it was a child, or someone carrying a large package, you would automatically

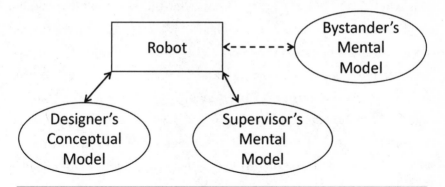

FIGURE 11: Diagram showing how the designer's conceptual model needs to be expanded to include a bystander. Bystanders need a passive mental model of robot behavior to achieve the shared goal of the two not interfering with each other.

give them a little more space to pass. Passing safely doesn't require you to know who that person is, where they are going, or why they are going there. In the same way, you don't need to know what exactly the robot is doing. Providing a bystander with a rich model of the robot is unnecessary, and besides, it would come at too high a cost in terms of cognitive load. Nor does the robot need to know exactly what you are doing and why. What if you had a nosy neighbor who frequently stopped you in the hallway at your apartment building to ask you where you were going, what route you were taking, how long you would be out, and what you were planning to eat for dinner? Or worse, what if your building's security guard started asking you all those things? These are details they should not care about. You might feel annoyed, or even alarmed, and say as little as possible and be on your way. You would likely try to avoid interaction with them in the future.

So only minimal clues are needed to allow robots and humans to pass each other safely on the street. But how can we provide bystanders with the right cues to predict robot behavior, so that they can adapt their own actions appropriately with the limited information they will have from observing them? And how do we design robots to accurately take in cues from people and adapt their behavior to avoid clashes?

This whole issue requires us to reframe our thinking about the design of robots. Instead of simply designing an automated system, we have to design a human-machine system.

We think about the people around us—bystanders to our lives— and consider them in our decision-making all the time. We hold the door for someone carrying heavy grocery bags, yield to children dashing into the street to grab a ball that's rolled away, give up our seat to an elderly person on the bus. Despite ideas to the contrary in science fiction, robots may never be capable of empathy, and they will not exhibit the same social considerations and courtesies that people show each other day in and day out unless we program it into them. They may not need to be perfect, though. It may be enough for these new social entities to simply seem like they are trying their best not to be a nuisance.

Let's take a simple example of merging into another lane when driving, breaking it down into the three levels of situational awareness—perceive, comprehend, and project. First, you perceive information about your own car (such as speed, position, and acceleration) and about the other vehicles around you (again, speed, position, and acceleration, but also proximity and type of vehicle). Then you move on to comprehension. If the driver of a car that is close to yours is texting, you comprehend that he may not be aware of you, and you project that he may not slow down when you merge. Or you perceive that a city bus is looming, but a space is open behind it. As a consequence, you may choose to delay merging even if there is enough space in front of the bus. Of course, at the same time you are perceiving other details—about the weather, about road conditions, and about other traffic changes beyond the cars immediately around you, such as car brake lights coming on in the distance ahead of you. This kind of implicit contextual information is usually not included in the conceptual models that designers develop for self-driving cars,

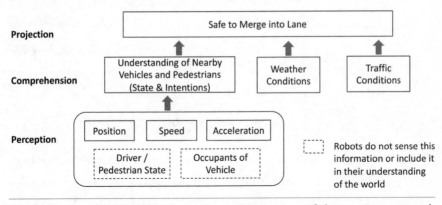

FIGURE 12: An example of how a person perceives aspects of the environment, such as the position of another vehicle and the state of the driver in the vehicle, and then uses that information during the comprehension stage to understand the situation and ultimately project whether it's safe to merge into another lane at a particular moment in time.

but it is nonetheless critical to safely interacting with other entities on the roads.

If bystanders are not included in the design of working robots, then as robot machines fill our roads, sidewalks, and skies they will become more than just annoyances; like the sidewalk delivery robots in San Francisco, they will become hazards, and their errors may result in real catastrophes. Thus they must be designed, at their core, to accommodate bystanders. Getting this right may in fact be our most significant new design challenge today.

Of course, in some cases, we do expect more from humans than just to stay out of our way. For example, if you were to drop some of your groceries while leaving the store, other patrons may pause to help you gather your items off the floor. Pausing to assist you does not help them get where they are going any faster—in fact, it does just the opposite. Nor do they do you this favor because they think they may see you again, and you might be able to help them in some way. The question is, should we try to develop robots, too, that do more than just stay out of our way?

We begin by focusing on minimizing interference between humans and robots, because getting this right is a prerequisite for better social interactions. More urgently, it can be a matter of physical or psychological safety.

IMPOLITE ROBOTS, IMPOLITE BYSTANDERS

People can implicitly understand the intentions of others around them without verbally communicating about it. We pause to let someone else speak, or slow down as we drive by construction workers. We accommodate people we don't know many times throughout the day without any explicit coordination or even conscious thought.

In each of these situations, many subtle cues help us infer the intentions of others and communicate our own. We make eye contact with other drivers, or hear something coming up behind us as we walk down the sidewalk and, turning around and seeing that it's a mom jogging with a stroller, move aside to give her and her baby a few extra inches to get by. We learn these social norms over many years, and not always through explicit training. They come so naturally to us that it is easy to forget just how much they organize our lives and enable us to efficiently function around each other. But just think how alienating it can be to visit a different country, with different norms. If you are from the United States and visit Tokyo, each time you enter a subway you have to remember whether you are supposed to walk down the stairs on the right side or the left. If you get it wrong, you are quickly reminded of that fact when a rush of people leaving the platform comes toward you.

Robots aren't yet designed to understand these social cues. Nor do they project their own intentions through such cues. If we want robots and human bystanders to be able to understand each other's intentions, we will have to make some new rules for them. These new rules will have to address two critical deficiencies in our mental models of robots, and vice versa:

- **The perception gap:** The gap in the cues and information we need to perceive other entities around us
- **The comprehension and projection gap:** The gap in the models and rules necessary to interpret those cues and accurately predict others' states and actions

There are two primary approaches to closing these gaps: expanding robot intelligence, and designing robots in a way that minimizes interference, so that human bystanders can more effectively negotiate the interaction.

DESIGNING ROBOTS THAT UNDERSTAND BYSTANDERS

Robot intelligence must be expanded with the following to make robots capable of properly behaving around bystanders:

- A broader understanding of human behavior beyond that of the supervisor, including a general understanding of the goals, actions, and situational context of bystanders
- An ability to sense and respond to subtle, implicit cues or explicit directions from people to minimize the time and effort it takes for us to communicate with the robots
- An ability to respond appropriately in a situation if interference occurs

The basic idea is to develop robots that can better understand the nuances of our everyday lives, and that can learn more effectively from fewer and shorter interactions with us. Roboticists and AI researchers naturally favor an approach that expands robot intelligence—it is an active area of research, with some recent laboratory successes.[5]

Until recently, in work settings such as hospitals, helper robots needed to be told exactly what to do and when to do it. When the robot's task is simple, and its schedule is very predictable, this works well.

For example, the Queen Elizabeth II Hospital in the United Kingdom employs a robot that cleans more than two hundred thousand square feet of floor per day.[6] It uses laser scanners and ultrasonic detectors to navigate, and if it encounters a human, it says, "Excuse me, I am cleaning," and maneuvers around him or her. But the robot relies on a human operator to give it an initial tour of its cleaning route, and then follows the same path over and over again. In some cases, robots of this type can become confused about how to pass by people, and remote human supervisors have to step in to directly control them. This approach is generally acceptable so far, but only because there are relatively few of these robots roaming the hospital's corridors at this point. Moreover, the robots can be scheduled to operate at night when there are fewer people walking around the hospital. Hospital workers have gotten used to the robots' behavior because it is so predictable: it follows predefined routes, and the humans have learned how to give it space to do its job.

Other helper robots in the hospital have to work more closely with people and must be more dynamic. For example, some robots are tasked with delivering food, medications, or linens. Their work schedule and the paths they travel change constantly as new patients arrive or current ones change rooms or are discharged. Nurse supervisors cannot be expected to explicitly command a fleet of robots step by step, in addition to their regular task of supervising seven to fifteen other nurses—it is just too burdensome. As a result, hospital staff members often reject these kinds of robots, because they feel like they might as well have a person do those jobs.[7] But if a robot could learn enough about the functioning of a hospital floor to jump in and help the nurses, then the upside could be tremendous. The Joint Commission for hospital accreditation in the United States found that 80 to 90 percent of sentinel events—that is, events resulting in death or near-death—in hospitals were due to human factors, including cognitive overload and failures of communication and teamwork.[8]

A robot assistant that could offer suggestions for even a portion of a nurse supervisor's decisions—where the nurse could accept suggestions rather than performing the mental calculations necessary to determine the best course of action on their own—could reduce the nurse's cognitive load and free him or her up to pay more attention to the many ambiguous and uncertain situations that nurses are trained to deal with, but that robots cannot do.

In our lab at MIT we tested new machine-learning techniques that enabled a hospital robot to learn to predict the comings and goings of doctors and nurses as well as the progress and needs of patients on the hospital floor.[9] This was achieved by having the robot spend just a few hours observing what took place on-site. It was quite a feat for the robot, given the vast number of decision points and possible futures in running a hospital floor. The robot watched nurses make decisions about which patients would be assigned to which rooms, as well as which nurses would be assigned to which patients, and merged that information with its data on the health status of the patients to predict what would happen next. The robot was right in its predictions up to 90 percent of the time, on average. It was able to achieve this level of performance by learning the way people do—through examples and counterexamples. The robot watched what actions the nurses took and didn't take, in the many situations that arose, and constructed hypotheses about the combination of factors that must be affecting the nurses' decisions.

This level of performance is very exciting, because it means that up to 90 percent of the time, the robot does not need to sit on the sidelines, waiting to be told what to do. Even when things change quickly, the robot can learn the rhythm and flow of our most difficult work environments and proactively offer help and services. But what happens in that other 10 percent of the time, when the robot predicts behavior incorrectly?

As AI researchers and developers, we see these as promising advances. But we are still a very long way from designing robots that are smart and sensitive enough to interact with bystanders as easily and naturally as people do.[10] People take many subtle cues into account about the people around them without even consciously thinking about it. Robots are not designed to look for this information, and their behavior suffers as a result.

In February 2016, for example, a Google autonomous car collided with the side of a city bus.[11] The autonomous car was driving in the right lane when it approached a sandbag in its way. It slowed down and began looking for an opportunity to merge into the left lane. A city bus, meanwhile, was approaching from behind in the left lane. There was enough space for the autonomous car to enter the left lane, the city bus began to slow down, and the autonomous vehicle took the opportunity to merge. The problem was that the city bus driver was not actually slowing down with the purpose of letting the autonomous car merge.

We all know that city bus drivers have a lot to manage in addition to driving: they are making sure all the passengers have paid, fielding questions from passengers, and looking for the next stop. Consequently, city buses commonly jerk between slowing down and accelerating, and so we typically give them extra space. The autonomous car did not have this same understanding of typical bus behavior. It applied the same logic and behavior that it would have applied to a car or a truck. Certainly, the autonomous car can't read a driver's body language when deciding whether to merge in front of a vehicle, regardless of vehicle type. So the Google car sideswiped the bus, damaging both vehicles. No one was hurt, but it was the first time a self-driving car was found to be at fault in a traffic accident.

A robot, whether on the road or in any other situation, cannot draw from a whole lifetime of accumulated experience the way a person can. The reason is that the robot does not sense and understand the world in the same way that a person does. Machine learning gives robots a shot at learning across a large set of "experiences," as defined by the data we explicitly share with it, but it is still not as flexible as we are when it comes to integrating its learning across a range of situations and contexts. The robot's model of us is drastically impoverished compared to the models people hold of one another. And it's not clear how much better we can expect robots to get at developing their "bystander model" of us.

In Donald Norman's example of the digital watch, he points out that a user has to press the buttons and observe what happens to learn how to use it. The user has to learn about the watch's functionality through a frustrating and time-consuming process of trial and error. Human behavior is impossibly more complex to a robot than a digital watch is to a human user. Robots' interactions with bystanders are brief and varied, and the types of social cues we rely on in human interactions can be very subtle. It takes experience and time to learn them, they change over time, and they can vary from one culture to the next.[12] Moreover, robots are not really members of our communities, so it's unlikely they will be able to build a deep rapport with people. And learning through trial and error comes at an unacceptable cost. There's clearly a cost-benefit trade-off to trying to make robots more and more like humans, and the sheer complexity of all the variation and subtlety in our daily interactions makes it unlikely that robots could ever become capable of understanding them even if we were to invest heavily in the effort. Fortunately, enabling robots to make better sense of us is only half the solution.

DESIGNING ROBOTS THAT BYSTANDERS UNDERSTAND

The other half of the solution is to minimize interference by designing robots in such a way that human bystanders can more effectively negotiate the interaction. With a human-centered approach, we can design the bystander interaction in a way that allows the human bystander to:

- **Understand the situation:** A human bystander should be able to quickly develop or access an appropriate mental model of how the robot will behave in order to determine whether intervention is necessary.

- **Understand how to communicate:** The human bystander should be able to communicate, implicitly or explicitly, to influence robot behavior and minimize interference.
- **Understand how to respond:** The human bystander should be able to take appropriate action in the situation if necessary.

It would be foolhardy at the present time to assume that robots can be relied upon to always completely understand the nuances of human behavior. People misread each other frequently, too, but at least we have rich means of observation and communication to quickly identify our errors and correct them. For example, we might stop and start in sync with another car at a four-way stop, with one or both of us confused about who has the right-of-way. Ultimately, one of us will wave the other through. But what would you do at a four-way stop if there was no driver at the wheel of the other car? Robots will have to be designed to address these types of problems—that is, to address the influence of bystanders' perceptions in an interaction, their comprehension of the events, and the projection gaps, the three components of situational awareness. We already have techniques for easing the difficulties associated with unpredictable human-human interactions. Because of large "Student Driver" signs, we can immediately identify student drivers on the road, for example. We take our cue from that to be sure to make eye contact, use signals, and communicate our intent. We need the same types of cues for our interactions with robots.

The city bus driver probably saw the Google autonomous car in the right lane slowing down for the obstacle. However, the driver may not have realized it was an autonomous vehicle, and, even if he had, would he have guessed that it would quickly merge into a tight space? The autonomous car's behavior was as unpredictable to the bus driver as the bus's behavior was to the autonomous car.

What is the language—verbal, aural, or physical—for a robot to signal just what a bystander needs to know? How do you say "Excuse me" to a robot? The way we communicate with robots doesn't need to be the same as in human-human interaction. We don't need a physical robot behind the wheel of an autonomous car to wave at us. But new modes of communication do need to exist, and they need to be appropriately designed into the system for humans and robots to be able to accommodate each other.

Designing robots that bystanders understand is made complicated by a number of factors, including the fact that people build understanding and trust slowly, bystanders' interactions with robots are brief, and the way a robot behaves can be unpredictable in new situations or when there are software updates. Designers and supervisors have the benefit of building their understanding and trust in the system through experience and practice. But what can a person learn about a robot in just a moment or two?

We trust automated systems when we believe they will help us achieve our goals in a situation in which we are uncertain or vulnerable.[13] When the goals of the supervisor are well defined and aligned with the robot's, the collaboration is more likely to be successful. A substantial amount of research has focused on what influences a supervisor's trust in automation, with the goal of designing systems that humans will trust—but that they will trust neither too little nor too much.[14]

Despite all the research, it is difficult to translate these lessons into the design of everyday robots. Calibration of trust takes time. In a 2017 study at MIT, we looked at a simulated scenario in which a person had to perform an independent task—controlling a dynamic vehicle (like an aircraft or a car)—while also tasked with continually checking the environment for potential threats to safety, such as a bad guy who might set off a car bomb or shoot the aircraft out of the sky. An intelligent agent was provided to help look for threats. The agent was not perfect, but it would try to alert the person if there was a moderate or high likelihood of a safety threat, so the driver or pilot could take appropriate evasive action.[15]

Imagine a similar scenario on your city streets. You are walking to work, and robots are zipping along on the sidewalk. Each robot has to decide whether to alert you as it sneaks up behind you to pass by. It does a calculation: How wide is the sidewalk? How likely is the pedestrian to lose his or her balance if startled? Might the pedestrian suddenly veer left or right? There are many robots, and the nuisance factor of them all alerting every pedestrian all the time is too much. So the robots make an imperfect calculation. If a robot guesses you might not hear it coming up, perhaps because you are wearing headphones, or it is moving so fast that it determines a collision might seriously hurt you, it will beep loudly and say "BEWARE!" But most of the time, they beep softly, perhaps with an unobtrusive jingle, just to let you know to look out. Sometimes, if the risk is perceived to be low, they pass silently. How many interactions does it take before you will instinctually know how to respond to these cues, whether to step out of the way or continue along, without it taking much mental energy or thought?

Our study says thirty-five to fifty interactions.

And the task becomes much harder for people even with small increases in the complexity of the robot's behavior. For example, in our study, when the robot had just two modes of interaction—silence and the "BEWARE" alert—it only took people ten to twenty interactions to calibrate their trust. A simple increase of one extra communication more than doubled the number of times people needed to interact with the robot before they understood how best to respond to it—for example, whether they should immediately heed the robot's warning and stop what they were doing to step aside, or pause to gather a bit more information about the robot with a quick glance over their shoulder.

Now imagine that every robot manufacturer designed their own unique alert. Some would beep, some would speak to indicate danger, and all with different underlying logic for when to alert a bystander. It's too much for any one person to keep track of.

Fifty interactions may be an acceptable number of times for improving your interaction with a teammate you will work with every day for the long term. Working together at first may be slow or awkward, but over time you will find a rhythm and begin working as partners. Bystanders do not have the luxury of time. They may cross paths with hundreds of robots, and if each one is communicating or acting differently, that's a problem. How many interactions would it take for the average person to learn all the different ways these robots may behave or respond to them, or to find the right response, beyond just getting out of their way?

Robots today are aliens to us, and just as robots lack a lifetime of experience to draw from in interacting with us, we lack experience with robots. We have trouble building our understanding of robots and our trust in robots because we don't have good mental models of robots to begin with. We drastically change our behavior when in the presence of robots. That means it's not good enough for a robot to just observe human-human interactions and then try to act as "humanlike" as possible.

We learned this lesson from a study we conducted to better understand a person's decision-making process when working with a team involving a robot, as compared to working with two other people.[16] We first invited teams of three people into the lab and studied them as they collaborated to assign tasks to members of the group, and then as they coordinated their actions to perform the assigned tasks. Unbeknownst to two of the people, the third member was a researcher and was tasked to perform the work exactly as our lab's robot would, taking exactly the same amount of time as the robot. All the members of the team were told in advance what each team member could and could not do and how long it would take each to do it (based on an initial screening test). When three people worked together, they allocated the work among themselves very effectively, almost perfectly evenly. However, we found that when we replaced the researcher with our robot, the people on the team behaved strangely. The human team members hoarded work from the robot and tried to actively decouple their work from the robot's. As a result, most of the mixed human-robot teams performed the tasks much less efficiently than the human-only teams. In both human-only and human-robot teams, the members were provided with the exact same information about their partners' capabilities, and the information was accurate. But people were not able to internalize the information about their robot partners in the same way they did for their human partners. In our view, we may never become sufficiently familiar with robots to fully acclimate to interacting and working with them. Robots are fundamentally different from people, or even some animals, such as dogs, which are capable of learning to read some of our body language.

When people are given information about other people—what they do well, whether they are likely to meet deadlines—they don't just blindly rely on that information. We use all our senses as well as previous experiences to critically analyze the data we've been given and come to our own conclusions about how much to rely on the information. Sometimes our mental models are flawed, leading to well-documented biases like those mentioned in the previous chapter, but more often than not they are helpful to us in our everyday lives.

But although you can tell a lot about a person even with first impressions, you can't tell much about a robot just by looking at it. The software governing what it understands about the world and how it should behave is designed and uploaded independently of its physical system. Software can be transferred from robot to robot, and it can be updated overnight. Robots' shapes, designs, and personalities can change, and even make huge leaps in functionality, between generations. And robot generations historically operate on much shorter time scales than human ones. You could be meeting a new robot every year that is designed to do the same task as the one before it, only slightly differently. In contrast, we can't yet do a brain transfer between people, or rewire a person's brain overnight to think and behave differently. Our behavior is deeply tied to our personal experiences in the real world and the resulting biases that affect our understanding of entities around us, all of which makes learning from our real-world interactions meaningful.

If nothing else, people are consistent over time, and there's little reason to believe that robots will ever be that consistent. The question to ask is how we can design for this problem. Making a robot tell us everything about what it could or couldn't do wouldn't be sufficient or enable us to work well with it. Directly giving a person this information, in an effort to close the perception gap, wouldn't mean the person could internalize the information and make good decisions about how to interact with the robot.

CLOSING THE GAP

What we need is a structured design process for closing the perception, comprehension, and projection gaps and enabling effective negotiation to take place between a bystander and a robot.

The goals of the bystander are different from those of the robot, and likely do not include the robot. For example, on the sidewalks of San Francisco, say a pedestrian wants to get where she is going efficiently and with a sense of security. The delivery robot is not acting with the pedestrian's psychological safety in mind, and so is not able to avoid interference in accordance with this goal. Additionally, the remote supervisor of the robot does not know the intentions of the bystander, either, and it is difficult for the supervisor to quickly gather that information. The robot's remote operator interface has a narrow field of view, and the subtle nuances of human facial expressions are hard to detect by that means. These challenges point to the need for more than human supervision of the robot as a solution to manage these potential conflicts.

The designers need to expand the goals of the robot and the supervisor to include *taking action to minimize interference with bystanders.*

Action actually has two stages: doing something (execution) and comparing what happened with what you wanted to happen (evaluation). Execution is based on your goals—that is, your overall intention, which you break down internally into a sequence of actions you think will achieve that overall intention. Evaluation begins with your perception of the world—which now includes the robot—and is interpreted based on your goals and intentions. In other words, you observe what your action achieved and compare it with what you intended in order to determine whether it met your expectations.[17]

Robots are merely a part of the physical world that we engage with. However, they pose special challenges to bystanders that make it more difficult to determine the correct action, thereby widening what Donald Norman called "the gulf of execution"—that is, the gulf between what you want to achieve in your interaction with some kind of device and the reality of all the steps you need to know how to do in order to achieve it. You just want to navigate your way around a robot, but what are all the steps you need to take to actually get around it? That gap is the gulf of execution you are facing in that situation. We have mental models for most of the objects that are present in our environment, and generally we understand how to interact with them based on our lifetime of experience with similar entities. San Francisco residents had not yet developed a sufficient understanding of the behavior of the sidewalk robots to be able to go about their day without feeling like the robots were a nuisance. Similarly, the intentions of the autonomous car were not knowable to—or easily influenced by—the bus driver.

Bystanders first need to know whether there is a robot present, then whether that robot will interfere with their tasks, and if so, what can be done to avoid a conflict. That makes for a pretty wide gulf of execution. Norman points out that the design of the system should allow for this gulf to be bridged with as little effort as possible by the user. The standard is even higher for the bystander, who, remember, is not gaining anything from this interaction with the robot. For the system to operate in our midst safely and effectively, bystanders must be able to understand the system's execution almost effortlessly. Expecting a person walking down the sidewalk to stop whatever they are doing to engage in a detailed interaction or negotiation with a robot passing by is a losing approach. The person will most likely go out of his or her way to avoid the bothersome robot or get it removed altogether—cue complaints to the mayor and board of supervisors.

Given that interference is inevitable, let's look at the potential ways in which we can bridge the gulf of execution, thus closing the gaps in perception, comprehension, and projection so that bystanders and robots can negotiate their interactions effectively.

In any interaction between a human bystander and a robot, the human is likely to have a few questions in mind. The first question is, "Will the robot interfere with what I'm doing?" The bystander's goals are different from the robot's, but the robot will enter the bystander's physical space regardless. Therefore the bystander needs to be able to predict whether the robot will interfere with his or her goals and planned actions in any way.

This challenge must be addressed through a *human-centered design approach*, in which we design the human interface with the robot so that the human bystander is able to quickly develop or access an appropriate passive model of how the robot will behave during their transient interaction. For example, the robot could have a way of signaling when it detects the presence of a person in its path, or when it plans to make a sharp turn, similar to a blinker. This signal would help bystanders predict the robot's path, and let them know whether the robot sees them and is planning a path around them. Bystanders need to know where the robot will be and when it will be there, and they need to know what the robot knows about them and the environment. They also need indicators of performance, so they can predict whether the robot will behave capably and meet their human expectations in the particular situation or context.

The next question is likely to be, "If the robot is interfering, what can I do about it?" If the robot and human paths are about to intersect, such that the robot is about to interfere with the bystander's actions, then there will need to be a way for the two to communicate with each other and adapt their behavior to alleviate the impact of the interference.

This challenge must be addressed through a joint *expansion of robot intelligence*, as well as, again, a human-centered design approach.

In this design approach, the robot's response should be based on the following:

- An understanding of situational human behavior, including short-term goals and actions
- A response to the bystander's implicit cues (such as a specific hand gesture) or explicit directions
- Identification of interference situations and selection of safe action

The human response would be informed by an interface that supports the bystander in perceiving the following:

- When the robot has identified a situation with potential interference
- Changes in the robot's future state, actions, and plans, such as changing direction or lanes to accommodate the bystander
- How a bystander can influence the actions of the robot

Once the human or the robot takes action to avoid interference, both need to have the capability to evaluate the effectiveness of that action. Usually, evaluation comes naturally to humans: if you chose one path of action, and it didn't work, you'll likely try something else next time; if what you did worked, you'll repeat it the next time you are in a similar situation. But robots pose special challenges to evaluations, in what Norman has called the "gulf of evaluation."

It can be difficult for us to evaluate our own actions in human-robot interactions, but it is even more difficult—or impossible—for robots to evaluate their actions. Today's robots cannot sense enough about people to fully comprehend the subtleties of human behavior. It's impossible for a robot to "figure out" whether you were satisfied with an interaction. When you eventually got around the robot and went on your way, were you happy, disgusted, or afraid? The robot does not know. Robots will also need to be able to factor in societal norms based on what they are perceiving in the world. Currently, they cannot do this, and these deficiencies make them unable to form an adequate understanding of the people around them and their responses.

The gulf of evaluation can only be bridged if robots are able to provide visible feedback to bystanders, and if we can provide sufficient feedback to robots to help them learn. We need to design robots in a way that allows bystanders to figure out how to respond to them quickly. Building trust is also a part of bridging this gap—and taking thirty-five to fifty interactions to figure out how to respond to a robot does not meet that minimal effort criteria. We need to find a way to make this easier for bystanders.

The next question for a bystander is, "Will the robot's behavior meet my expectations?" When we interact with other humans, we rely on cues about whether our expectations for their behavior will be met—based on, for example, eye contact, simple words, or some other form of body language. Robots must be able to provide us with similar types of cues when they intersect with us, to enable us to perform our mental evaluation.

This challenge, again, must be addressed through a human-centered design approach. We should be able to use cues to develop or access an appropriate mental model of how the robot will respond to our actions. These cues would be based on the following:

- Feedback upon recognition of the bystander's implicit or explicit directions to the robot
- Indicators of the robot's understanding of human behavior, actions, and situational context

Bridging the gulf of execution and the gulf of evaluation for bystanders explicitly during the design of the robot will help to close the gap that exists between robots and the people they will inevitably interfere with as they enter more directly into our everyday lives.

THE CATASTROPHIC COST OF THE THREE-BODY PROBLEM

The three-body problem isn't only about minimizing the nuisance factor. We are only now getting early glimpses of the catastrophic costs of turning a blind eye to this problem. In Tuscola County, Michigan, on a rainy afternoon in August 2017, a man was heading westbound on a highway when his car began to hydroplane.[18] He lost control of the vehicle, crossed the centerline, and hit an oncoming pickup truck. Emergency crews were dispatched to the scene to free him from his car; he was trapped and severely injured. A medical helicopter, Flight Care, was requested, and in the meantime on the ground, the EMTs went to work, carefully extracting him using the Jaws of Life.

As the Flight Care pilot approached, he quickly identified the crash scene and initiated the landing sequence. He began to descend. During a scan, the pilot caught a flash out of the corner of his eye—his heart raced, abort! He barely averted the growing threat, a recreational drone circling over the crash scene to capture video of what was happening. The emergency rescue was delayed until a new landing plan could be developed, and the driver passed away later that day from his injuries.

The pilot couldn't continue descending because he could not be sure what the drone was going to do, and he had no way to communicate with it. He was merely a bystander, unable to direct or predict the drone's behavior.

Clearly, the goals of the helicopter pilot and the goals of the drone operator were not aligned. The pilot of the rescue helicopter wanted to safely land the helicopter as close to the accident as possible, and as quickly as possible. The pilot was not thinking about avoiding drones as a part of this goal. But the supervisor of the robot did not know the intentions of the bystanders, and it was difficult for him or her to quickly gather the appropriate information through the narrow field of view of the remote-control interface. The supervisor of the drone was looking through the video feed and maneuvering the vehicle to get a better angle on the accident, and may not have been aware of the rescue helicopter above the drone.

The rescue helicopter pilot talked to air traffic control and listened on the radio to avoid any conflicts with manned aircraft in the area. On the ground, emergency personnel prepared an area for the helicopter to land and kept people out of the way, minimizing bystanders in the area. But the intentions of the recreational drone were not knowable to the pilot and not easily influenced.

Unfortunately, this was not an isolated incident. From 2014 to 2016, about *650 cases of near-misses between drones and other aircraft*, including commercial airplanes, helicopters, and firefighting aircraft, were reported to the Federal Aviation Administration.[19] Innocent operators, using these flying robots to capture video, race with their friends, or try out new features, are unintentionally interfering with important activities.

These drones are remotely piloted today and have little or no autonomy, but they are the canary in the coal mine. They signal a larger threat to come as we have more and more robots in our skies, and soon, up close and personal on our streets and sidewalks.

CHAPTER FIVE

Robots Don't Have to Be Cute

WE TEND TO THINK ABOUT ROBOTS IN TERMS OF HOW MUCH joy and entertainment they bring. We buy an Alexa not only to play our favorite music, but also to add character to our homes. We delight at her preprogrammed banter, her jokes, and her animal sounds. People personify their Roombas and choose smart home devices that blend with their decor. We give our devices names as if they were pets, and customize their voices. The overwhelming sense is that what we want from robots is for them to be relatable. We want them to be pliable pseudo-people.

In turn, robots today are typically designed with special attention to aesthetics and character. When news stories about robots go viral, it's because the robots have been made to look more like people. They mimic our facial expressions and act friendly. We want them to have personalities. Indeed, a great deal of attention has been devoted to developing robots that can elicit engagement from their users and connect with them on an emotional level. The companies that develop these tools likely feel that anthropomorphizing their products will help create attachment to their brand. There is a whole new field of technology design that aims to optimize the user's emotions, attitudes, and responses before, during, and after using a system, product, or service. It's called *user experience* (UX), and for most businesses the goal of UX development is to attract and keep loyal customers.

But as robots enter our everyday lives, we need more from them than entertainment. More and more, we don't want them to simply delight us, we want them to help us, and we need to understand them. As robots weave in and out of traffic, handle our medication, and zip by our toes to deliver pizzas, it won't really matter whether we're having fun with them. Developers of new technology will have to confront the complexity of our everyday world and design ways of dealing with it into their products. We will all inevitably make mistakes in these interactions, even where lives are at stake, and it's only through designing a proper human-robot partnership that we will be able to identify these mistakes and compensate for them.

The stakes for the design of most consumer electronics are now fairly low. If your smartphone fails, most likely no one will get hurt. So designers focus on providing the best experience for the most common situations. Problems that arise in only rare circumstances are tolerated, and the assumption is that most problems can be solved by rebooting the device. If that fails, you just have to figure it out, perhaps with the help of a tech-savvy friend. It's simply not the point of most consumer technologies to be resilient against all possible failures, and it's not worth the effort for companies to prevent all failures. A user, after all, is usually willing to overlook an occasional software glitch, as long as the overall experience is enjoyable and the device seems more useful than what the competition has on offer. This just isn't the same for safety-critical systems: a blue screen of death on the highway in a self-driving car could mean a catastrophic accident.

So the goal in UX is to elicit a positive emotional response from the user, and the best way to do that is to focus on the artistic aspects of the system. Give it a "personality," make it sleek and gamelike. Emphasize product branding. Consider trapping users within the system by hoarding their data or otherwise making it hard to transition to a competing product. And then, at some point, stop sending software and security updates. The planned obsolescence of the product forces the user back into the sales cycle. The ultimate design goal of most consumer electronics is to make people buy more of them, which results in short time scales between generations. And every time you purchase the newest version of the product, you have to restart the learning process.

These design goals will not be sufficient for the new class of social working robots we will be encountering more and more in our daily lives. Take, for example, the first BMW iDrive. BMW was on the cutting edge of the movement to introduce high-tech infotainment systems to cars. In 2002, the company debuted the iDrive. The engineers tried to make it fun and sleek, but that wasn't enough. Just as in the introduction of new generations of aircraft automation, this first interactive infotainment system brought about unexpected safety concerns—so many of them, in fact, that early versions of the system were dubbed "iCrash."[1]

The first iDrive provided users with flexibility to customize the display to fit their preferences. There were approximately seven hundred variables for the user to customize and reconfigure.[2] Imagine how distracting it was to modify the placement of features or the color of buttons on the screen while stopped at a red light. It created unnecessary complexity for users, because there was too much to learn. The extensive features of the infotainment system and the many ways to customize it were overwhelming. As drivers became consumed by fiddling with the interface, their focus narrowed, and things became dangerous. Drivers began to miss important cues about the road or other cars. This is why user customization is a bad idea for safety-critical systems. Instead, designers need to determine an optimal setup for the controls from the beginning with safety in mind. In this case, commonly used features needed to be more easily accessible to the driver. A single button to turn the air-conditioning up and down or change the radio station should not be hidden beneath a complex tree of menu options.

The physical layout of the first iDrive system was problematic. The design introduced a central control architecture with a digital screen and single controller, a trackball. But the display and controller were physically separated, with the screen in the central front-facing panel, and the controller on the central console between the two front seats. Most other infotainment systems had required the driver to press buttons near or on the display screen itself. The physical separation between the screen and the input device presented a mental hurdle, as drivers had to fiddle with the trackball in one location and watch the screen in another. Also, removing the physical buttons eliminated the muscle memory most of us have developed in our own cars. We reach over and grab the knob to turn down the air-conditioning fan without even taking our eyes off the road. This isn't possible with a digital screen: the driver has to look away from the road to adjust the air-conditioning or radio.

Finally, the first iDrive used a *deep menu structure*, which required the user to click through many menu options to access specific functions. Specific functions that users would want were buried deep within a series of options. A *broad menu*, separating functions into individual controls that could be accessed directly—such as knobs or dials—would have been better. The broad menu design is the choice for most cockpits, because it allows pilots to activate specific functions with a single button press. The pilot is physically surrounded by an entire set of menu options and can quickly activate any one of them at a moment's notice. Broad menus do require more physical real estate for the knobs and dials, and they might require the user to know more about the system, depending on how many menu options there are. They may look more complicated, but in fact they make it easier to select options quickly. The right solution for working robots, as we will see, often blends both approaches.

IN SOME WAYS, NEW ROBOTS HAVE MORE ADVANCED INTELLIGENCE THAN even commercial airliners. Advances in AI and deep learning made within the past few years allow robots to independently conduct search-and-rescue missions on forest hiking trails just as well as humans, for example.[3] But they are really just one component of the new breed of search-and-rescue teams, which are made up of both robots and humans. Even as robots become more capable, for the foreseeable future what we'll see more and more is humans and robots working together. Meanwhile, the gap between a robot that merely delights and a robot that does real work for us is widening. As that gap widens, safety issues become ever more important. While an infotainment system presents a distraction to drivers, it does not directly control any safety-critical aspect of driving. Nor is it considered an intelligent system by any stretch. But as we see more intelligent, autonomous technology applied to more and more areas of the modern world, such as operating heavy, fast-moving, and potentially dangerous pieces of equipment, there will be many more conflicts. For these applications, we care less about the design delighting us; we will want a design that will ensure our overall safety, whether we're users or bystanders.

It's unlikely that companies will be interested in abandoning the UX approaches that make their products so profitable. But they'll need to make calculated design choices to create a new era of human-robot partnership while still delighting customers. The good news is that the two goals are not completely at odds with each other. There is an important subtlety at work here. A renowned researcher of teams, J. Richard Hackman, has observed that it's not important for team members to actually like each other: teams made up of team members who like each other perform no better than teams made up of team members who don't like each other. Following a team performance, a team member's happiness depends more on whether the team was successful than on whether the team members like each other. This is further evidence that it's more important for robots to help us be successful at our tasks than it is for us to like our robot teammates. Our overall satisfaction with a robot as a product will come more from how productive we are together than from how much emotional attachment we feel toward it.[4] Consumers may find some necessary features annoying or distracting in normal conditions; perhaps they'll prefer at times to ignore their helpful robots and get back to the devices that delight them, and will start text-messaging, reading, or playing games. But once such features help them do their jobs—or control a malfunctioning robot on the sidewalk—their attitudes will change.

This doesn't mean we can simply turn the world into something like an airplane cockpit as these working robots hit the streets and sidewalks. We aren't pilots, and don't have the same luxury that pilots have to undergo training. Therefore, working robots will need to be based on mental models that come naturally to everyday users. To this end, designers can still leverage UX design, and look to the research that has been done in adjacent consumer products, such as mobile devices and automotive standards. But even with a strong adherence to UX design goals, focused on delighting both users and bystanders, they can add a complementary focus on safety and productivity.

Because UX practitioners today are supposed to capture the hearts of users, they spend a lot of time trying to get to know them. They create "user personas" to embody the user population as a single, tangible person, for example, a technique that enables a design team to empathize with the people in their target audience and try to predict their needs and preferences. They brainstorm new concepts, and graphic designers make sure the aesthetics are compelling. They

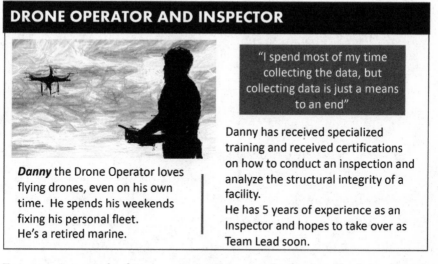

DRONE OPERATOR AND INSPECTOR

"I spend most of my time collecting the data, but collecting data is just a means to an end"

Danny the Drone Operator loves flying drones, even on his own time. He spends his weekends fixing his personal fleet. He's a retired marine.

Danny has received specialized training and received certifications on how to conduct an inspection and analyze the structural integrity of a facility.
He has 5 years of experience as an Inspector and hopes to take over as Team Lead soon.

FIGURE 13: An example of a user persona. This one is for the user of a drone that collects data for the inspection of industrial infrastructure to assess structural integrity.

conduct testing to give users a chance to try out new designs and give their feedback. Testers often see multiple versions of the same product, with designers and market researchers hoping to find out which one elicits the most positive emotional response.

There are also more quantitative and formal approaches to design. Designers might employ metrics based on usability principles, for example, in which real users try out the system.[5] Researchers then analyze their interactions with the system, or teams of expert evaluators come in and score the results. More thorough experimentation across a broad spectrum of situations would be useful. A system could be tested in situations ranging from normal, expected use to highly stressful scenarios, with sample sizes large enough to draw meaningful conclusions about the system's effectiveness. But this kind of research is usually well outside of the budgetary constraints and expertise of most design teams today. It is time-consuming and therefore not as widely used in commercial practice as in more industrial applications. Most teams revert to the fallback: building products that people seem to like.

Let's look at two examples from leaders in UX design. One is from Carnegie Mellon University, which has the world's longest-running master's degree program in human-computer interaction (HCI). Having hired and interacted with many graduates of this program, we can attest to its strength and impact. The other was developed at Google, an obvious leader in delighting users and building products that influence how we run our everyday lives.

The gap between UX design processes and the design needs for safety-critical systems can be traced, in part, to the training of HCI designers. For example, Carnegie Mellon provides rigorous and comprehensive training in systematic approaches to design and evaluation. Students get hands-on practice in the design process through interactive design studios and projects. They build skills and expertise in user interfaces, sensors, control, and ubiquitous computing (that is, computer technology that shows up basically everywhere and is pervasive; it is embedded into everyday objects and becomes a constant feature of our connected lives). They learn computer programming to build skills in prototyping; and they practice communication, writing, and team-based skills, including conflict management.

However, courses on human factors engineering and psychology (cognitive psychology, perception, and so on) are electives. The fundamentals of how people develop situational awareness and make decisions during stressful or uncertain times are considered secondary, optional concerns, because the primary focus is to make fun technology.[6]

This gap in training is evident in even the best consumer design practices. A developer at Google invented a process called *design sprint*, which was later described in a best-selling book.[7] In this iterative approach, designers work through an end-to-end design process quickly over the span of a few days. They improve their design in the next round based on initial results gathered during the sprint.

Each sprint takes one week and comprises the following five phases, each lasting one day:

1. **Map**—Marketing people, senior managers, designers, and sponsors come together to share ("map") knowledge about the design problem, envision potential solutions, and identify metrics to understand the impact of a solution. The team develops a set of "user personas" to better understand the characteristics of the user throughout the rest of the design process.
2. **Sketch**—The team works to develop rough sketches of specific design solutions.
3. **Decide**—The ideas generated during sketch day are compared and assessed against the objectives, abilities, resources, and users outlined during the map phase. Issues such as budget, technical capacity, business drivers, and users are considered. Once the early concepts are narrowed down to one or more contenders, the group creates a storyboard of the ideas.
4. **Prototype**—A rapid prototype is created for each idea that can be tested by users.
5. **Test**—Finally, six to twenty users are brought in to perform playtesting with the prototype(s). Feedback during the playtesting is used to iterate and improve on the design.

User interests and business objectives are considered equally throughout the process. The team gathers input from only a small population of users, which makes it hard to quantify task performance with statistics. Testing with a larger number of people would require more time and cost more, and when speed to market and profits are a concern, time and cost are important. This design approach favors producing systems that users say they prefer, or that they in fact preferentially use. It also prioritizes business objectives over optimal augmentation of human decision-making and task performance.[8] Of course, user preferences are notoriously fickle—perhaps that has something to do with the seemingly endless updates and "improvements" users of a product are subjected to.

The case is different for industrial systems, where safety is the primary goal. Here, designers want to compensate for human weaknesses, and user experience is secondary. Human psychology informs every aspect of these projects. The users of these systems do not need to like the systems, but they absolutely need to be able to operate them safely and predictably across a wide variety of scenarios, including, especially, in failure conditions. And just as the teaming research shows, users end up liking their robot teammates with this approach in the end because they help them achieve better performance.

The design team in this case is composed of experts in engineering fields dedicated to improving relationships between people and machines. They consider the physical and psychological characteristics of people in the design of their systems and environments by studying human factors engineering, engineering psychology, and cognitive science. These engineers are trained to understand expert decision-making and to analyze situational awareness. They want to know how people's cognitive processes change under different conditions, such as stress and boredom, and the role that intuition and biases play in human judgment.

The design team studies all the tasks required to operate the system; every detail, down to the placement and color of buttons and dials, is explicitly designed to optimize human decision-making and performance when operating the complex automation. For example, how far a switch should be from the pilot in the cockpit isn't just chosen randomly; it is modeled and intentionally selected based on the results of study and research, because the wrong design choice could result in delayed execution of a task. If that task is part of a time-critical activity, such as responding to a failed engine during takeoff or landing, that delay could be a life-or-death matter. Can you imagine if a pilot had to reach in one direction for some of the switches and buttons required for this task, and then reach in the other to complete the task, all while

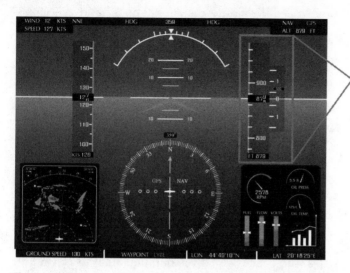

Analog tape of aircraft's altitude that has been digitized

FIGURE 14: Diagram showing an example of how analog displays have been converted into digital displays in modern airline cockpits to support the pilot in monitoring the direction and rate of change of the aircraft's position and speed with little cognitive effort.

experiencing high gravitational forces and the stress of saving the aircraft? The designers think through these issues.

These designers also pay a great deal of attention to how information is represented, so that users can draw the right conclusions in a timely manner. For example, in digital cockpits, analog displays, such as dials and tapes, are no longer necessary, as the value can now be provided to the pilot as a digital display, like the kind on a digital alarm clock. However, it is still much easier for a pilot to gauge how fast the altitude, say, is changing by watching a needle on a gauge move, as compared to watching numbers change on a digital display. A moving needle provides a much better sense of the direction and rate of change for a quantity such as speed or altitude and can be interpreted quickly with little effort. Digital numeric displays are reserved for quantities that must have a precise value, and for which the direction and rate are not as important. Presenting the right information, at the right time, in the right manner, to the user is not just a matter of good design, it is a matter of safety.

This careful process of designing displays, knobs, and dials may appear on first look to be a narrow, niche view of interaction with a system, maybe too focused on the details to be useful across the broad range of working robots we will see. But this perception is incorrect. Human factors engineers, engineering psychologists, and cognitive scientists create a good blueprint for designing collaborative robots because they are specifically trained to design human-machine systems. They understand that seemingly small factors in human-automation interaction may have cascading effects on a person's ability to do her job. For example, in designing a smartphone app for soldiers to help them navigate unknown terrain, communicate with their teammates, and plan out activities, the process involves detailed analysis of the full set of cognitive and physical tasks that soldiers may need to perform during each part of their mission, both for common operations and for a comprehensive set of error situations. The app needs to achieve a number of specific goals. It has to present detailed information to the soldier clearly, so the manner in which this information is conveyed is important (e.g., visual, auditory). It has to account for the physical and cognitive tasks involved for the soldier to understand the situation, make decisions, and act. And all of this has to be optimized across the automation that is programmed into the app. It's not just to optimize the soldier's interactions with the interface's menus and buttons, but to ensure that the soldier performs well in all the tasks necessary to carry out the mission. These soldiers, making numerous crucial decisions as they work together, certainly need the app to help them do their jobs, and don't care about the "delight" factor.

Rigorously analyzing how an operator interacts with a robot, and understanding the potential impacts on an operator's situational awareness, workload, and performance, is no small undertaking. Researchers must design detailed evaluations that assess quantitative performance metrics through extensive experimentation with real users. They have to introduce a range of failure conditions to see how they will affect the user's cognitive load, and examine how users perform when they have to multitask, and their attention is divided. They gather extensive quantitative data on system performance and human performance, statistically analyze it, and then use that data to refine and improve the design.

This is a costly and time-intensive endeavor, especially when the system is as complex as a commercial airliner or a system controlling operations on a factory floor. And that partially explains why these systems are so expensive to develop and produce. The cost of developing a new commercial airliner, for example, is usually more than $10 billion.[9] The high cost is due to the enormous design and evaluation effort that is required, as well as to the fact that many backup and redundant systems have to be built into the airliner to address all the failure scenarios that surfaced during design and development. The other hidden cost is the training of the operators. But when designing a system that will be responsible for the safety of hundreds of thousands of people, the high cost is appropriate. And it's appropriate in many other types of situations—for example, in the operating room, where a medical robot has to be almost unbelievably precise to aid in surgeries; in factories, where workers need to be able to monitor and interact with powerful and dangerous equipment safely; or even in weapons systems, where achieving the nation's objectives depends on these factors.

We face a significant challenge, then, in building safety-critical working robots: people will not be willing to pay as much for an errand-running robot as they would for an aircraft. Nor will they be willing to undergo ten thousand hours of training before we start using them. Furthermore, new working robots will be operating in environments that are less controlled than a cockpit, an operating room, or a factory, and less amenable to our traditional processes for certifying their safety. So how exactly will our procedures for building and testing robots for use on our streets and in our homes have to diverge from standard industrial design processes to address the complexities of this new class of human-machine system?

Besides, humans are far from perfect, so how perfect do robots need to be, anyway? For example, we don't expect people to be perfect drivers. Massachusetts requires only fifty hours of instruction for teen drivers, and then we outfit cars with safety technology, encourage defensive driving, and structure society with speed limits and other rules. We feel comfortable enough unleashing teens on our roads, knowing they'll learn and improve with experience. We accept a certain level of risk. So what kind of reliability and oversight should we expect of robots once they are deployed, whether they're on our streets, in the workplace, in the mall, or anywhere else?

In addition, how can UX approaches be integrated with other approaches to ensure that user experience objectives, business objectives, and safety objectives are all met? In fact, even industrial systems stand to benefit from a hybrid approach to design. Industrial robots meet safety standards, but certainly they do not "delight" users the way consumer products must. Could they delight as well, without harming safety—or perhaps even enhance safety by being delightful? Take the user interfaces and controls of bomb disposal robots used today. They work well for military applications, but usually there's nothing particularly delightful about them. One company that manufactured bomb disposal robots, however, gained market traction by incorporating UX concepts: it switched from the standard interface, shown in figure 15, to using two game-style hand controllers, because they found that these controllers were more intuitive to the operators, who tended to play video games in their free time.[10] If we weave UX techniques together with design techniques from industrial systems in this manner, perhaps we can take the best from both worlds to create highly effective intelligent robots that work well with us in our everyday lives. Sometimes, an interface that delights the user may also provide improved performance.

Interacting with an intelligent robot is fundamentally different from interacting with an interface that supports automation, such as an automated flight system, because intelligent robots act independently. A

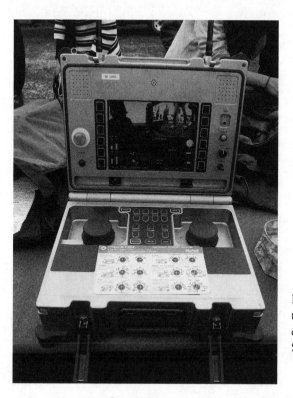

FIGURE 15: PackBot user controls for explosive ordnance disposal. *Source:* viper-zero / Shutterstock.com.

pilot and her plane always want the same thing. That's not the case with a delivery bot working on the street and the people around it. Intelligent robots have their own goals, intentions, and logic for how to accomplish those goals. Their behavior is likely to change over time as they learn and update their models of people and the world. Thus they will be much more complex than other forms of automation, and their supervisors must also expand their mental models of how the robot works and what it's capable of.

The military and civil defense industries were among the first to invest in the use of intelligent robots. The first uses included disaster response and bomb disposal, because such work is often very dangerous for people to carry out. Although we're only beginning to scratch the surface of what we'll need to know to design the new human-robot collaborations that will emerge, the early research from labs experimenting with the expanded use of intelligent robots has given us insights and lessons to build on. Some are old lessons that we had to relearn the hard way; others are new challenges unlike anything we've seen before.

The Robotics Challenge, sponsored by the Defense Advanced Research Projects Agency (DARPA), took place from 2012 through 2015 and brought together the world's top robotics researchers. Participating teams from different universities, institutes, labs, and other institutions competed in fielding intelligent robots that responded to simulated human-made and natural disasters. In the final competition, MIT's intelligent robot was eliminated from competition after failing to gracefully transition from one mode to another. The failure was due to the operator's mode confusion, a phenomenon identified and addressed in the design of cockpit automation.[11] The robot was tasked with driving a vehicle to a disaster site, getting out of the vehicle, and operating a saw to break down a wall. MIT's team was leading in the final competition until the robot took a dramatic fall exiting its vehicle.

The robot's intelligent software did not transition as planned from driving to walking mode, and the supervisor did not take the appropriate action to correct the error. As the robot stood up out of the car, its right foot continued to "pump" the ground, as if it were pushing an accelerator pedal. The five-foot nine-inch, 330-pound robot knocked itself off balance and fell to the ground.

MIT's robot could drive, walk, and saw through walls perfectly. But each of those tasks was governed by a different mode of operation. In driving mode, the robot articulated its ankle to press the vehicle's accelerator pedal. In walking mode, it moved its hip, knee, and ankle to produce a walking motion. In each mode, the supervisor was tasked with monitoring a different set of parameters to ensure proper functioning of the robot, with only a few "knobs and dials" to adjust the robot's actions and ensure success of the task. This is a very common approach to designing automation, and it is highly effective. It simplifies the tasks at hand for both the robot and the supervisor. However, the transitions become crucial as well. If the robot or supervisor does not gracefully manage a transition between two of the modes at the right time, the result can be an unrecoverable, catastrophic failure. MIT's robot could do many tasks very well, but it was unable to switch between those tasks.

In the design of industrial systems, we can overcome these pitfalls by providing the supervisor with the right information to support the three levels of situational awareness we learned about earlier—perception, comprehension, and projection. In the case of MIT's robot, the user could not see which mode the robot was in. The lesson was that if there are crucial transition points, then the designers have to make sure the supervisor can tell when the transition is underway and completed. Information and communication about the mode or transition point need to be prominent to the user. It may have only taken a simple push of a button, to turn on walking mode, for the robot to exit the vehicle successfully. Such transition points need to be absolutely clear to the operator. It was such a critical juncture for the MIT robot that the designers might have chosen to make the mode change a specially designed task initiated by the user and actively monitored by the robot. Such an approach would at the very least have ensured that the user maintained situational awareness about the robot's mode. Passively monitoring a series of mode changes can be challenging for anyone. But because we know it can be easy to miss a mode transition, we need to design the solution into the system.

This kind of challenge represents more than an interface design problem. It requires an understanding of the interdependence between person and robot. Much as in any partnership, the human-robot partnership develops through a series of actions and responses. But rather than letting interdependencies emerge as an unforeseen consequence of design, we need to design for the interdependence explicitly.[12] And to create these kinds of interdependent systems intentionally, we need to focus on three specific relations: observability, predictability, and directability.[13]

Observability involves designing the robot so that it can make pertinent aspects of its status, as well as its knowledge of the team, task, and environment, observable to others. This includes aspects such as the mode of the robot, its operational constraints, its objectives, and its developing understanding of its environment. There are display considerations, such as when and how the system's state is conveyed to the user. But there are also robot-behavior design considerations, such as what the system can understand about its own state, about its progress through the task, and about the larger context of its environment and other people. These factors influence how and what the system can communicate to the user, and incorporating them into the system's interface is a design challenge. But we have evidence from studies of military systems that the effort to do it pays off.[14]

Designing for *predictability* pertains both to robot behavior and to the robot's interfaces. The interfaces support the operator in projecting what the robot will do, and then in using this information to determine his or her own actions. Again, there are display considerations—how and when the system conveys its reasoning and projections to the user—and there are robot behavior considerations, such as what the system can understand about its own potential limitations and likelihood of success and failure.

Situation Awareness–based Agent Transparency (SAT), developed by the US Army Research Laboratory to guide the design of interactions between intelligent agents and human supervisors, is one approach to designing for the observability and predictability of a system. The SAT model distinguishes three independent levels and types of information that an intelligent agent must convey about its own decision-making process. By conveying this information, the agent enhances a supervisor's understanding of the system and situation, which in turn leads to better teamwork.[15]

For Level 1, the robot needs to convey its current status, goals, intentions, and proposed actions. For Level 2, it needs to convey its reasoning process. This includes its overall objectives and its beliefs about the world, given its current information and the constraints affecting it in the environment. Finally, for Level 3, the robot must convey its projections and predictions about the future, including the likelihood of it successfully completing its intended actions, any uncertainty pertaining to its ability to complete the actions, and its level of interdependence on others.

Designing for *directability* involves analyzing the ability of both the robot and the supervisor to influence each other. Using approaches from industrial engineering, tasks are allocated between person and machine, creating interdependencies. As the robot becomes more intelligent, and is able to perceive, decide, and act somewhat independently of people, these interdependencies become more complex, and it can be harder to tell how the behavior of one influences (or is influenced by) the other. Directability relates to how the entities are directing each other's actions and how.

Everyday objects are designed with affordances to help us readily perceive the manner in which we can productively act on or use the object. In other words, when we look at an object, we can see what it's for and how to use it. Think again of the example of scissors: it's easy to see how to hold them, and what they'll do for us when we move them in our hand. Another example is a doorknob, which is obviously something you grab to pull open a door. We will have to design automation affordances for intelligent robots, too, so that people can readily perceive how they can productively influence the system. These affordances will create means for supervisors, or even bystanders, to quickly infer what actions they can take to influence the robot, and what kind of response they can expect from the robot. All of this is essential to providing directability for the user. (We will explore the design of automation affordances in chapter 6.)

The early lessons we've learned from military and industrial systems are useful, but there is one critical difference between these systems and the new intelligent robots that will roam our world. The supervisors of intelligent military robots all have very extensive training for operating and programming them. They also build up extensive experience performing tasks in an operational environment. The military operators of the bomb disposal robot had, on average, over two hundred hours of experience working with military robots.[16] The supervisors of MIT's robots at DARPA's Robotics Challenge had designed and built the robot's intelligence system themselves, and had spent hundreds of hours operating it before the competition—you can't get more expert than that! We do not have the luxury of training everyone for hundreds of hours before they ask their delivery robot to drop off their dry cleaning. And there is no training at all for the many people who will share the sidewalks with these systems. There is still an enormous gap to bridge as we move from designing for experts to designing for ordinary users.

There is yet another design gap that isn't being fully addressed today, whether in the UX community or in industrial systems design. User interaction with a robot is often designed separately from the behavior and reasoning of the intelligent system. A dedicated team of roboticists and artificial intelligence experts focus on an elegant design of the robot intelligence, but this often results in a robot that doesn't actually understand humans. Moreover, often they just aim to re-create capabilities that humans already excel at, while completely missing capabilities that humans struggle with. As a result, they create a system that doesn't actually improve the overall performance of a task. The user may struggle with a part of the task that the robot designers didn't think of, such as attention allocation. A human factors engineer or UX team may be brought in separately to design an optimal user interface for the robot, but only after the robot's capabilities and intelligence are already set. There is no doubt that at this point poor UI design can render an intelligent robot unintelligible to people. And also that excellent interaction design can make a less capable and intelligent robot maximally useful.

However, once the behavior and reasoning capabilities of the robot are set, UI design is superficial—merely a means to achieving the capabilities already designed into the robot intelligence. Even the best UI cannot make up for the problem when the robot's basic objectives are faulty and its true potential has been missed.

There is a much more extensive design space available for engineers who set out to design a robot that is truly able to collaborate with humans on a task, rather than just being able to complete a task independently. When the tasks of the user and the robot are co-designed, and the engineers understand how the robot's intelligence can contribute to collaboration at a deeper level, there is an opportunity to create a real partnership. We can only design good human-robot teams if the roles of the users and the robots are designed as parts of a whole. That way, the robot's capabilities can be designed to augment the human's capabilities, and the human's tasks can shift to supporting the robot's weaknesses, as any good partnership requires.

For example, many self-driving cars are designed to follow cues such as lane markings. The user's role therefore de-emphasizes some of the standard aspects of driving. Instead, users should focus on overseeing the robot's job, such as watching out for problems with the lane markings, which may arise when there is construction or on worn-out roads. The supervisor will need to quickly access the correct information about lane centering and know precisely what to do to overcome any lane-placement errors when they arise. So the engineers need to design the car in a way that makes this easy to do. In the Robotics Challenge example, the roboticists hadn't ensured that the supervisor would be able to visualize the transition between modes and intervene at those moments.

We can address this problem in two different ways—indeed, we can use both ways, and they should be integrated with each other. One approach involves better visualization and interface design, so the supervisor is able to get information quickly and respond; the other involves better design of the robot's intelligence. Approaching these two paths separately can lead to poor human-machine team performance. Many studies, for example, have found that nurses in hospitals quickly learn to ignore constant auditory alarms at the nurse station.[17] UI design alone cannot fix the problem of having too many false alarms that drown out the real problems. Instead, the machine needs to be able to process the incoming data more effectively, and thus surface alarms only when they are useful to the nurses.

Similarly, a sidewalk robot that constantly alerts people as it approaches them would just add to everyone's overall stress levels. Instead of responding appropriately to the robot, we would just want to tune it out. On the other hand, a robot that could recognize and understand important characteristics of pedestrians—whether they are walking with a cane, or might be an unpredictable child, for example—and sound an alarm only when one is warranted, may be much safer and more useful. Such possibilities open up new opportunities for design as well as for developing more positive human-robot interactions. These considerations can be made apparent early in the design process. We are right at a point where the role of robots is expanding, and the potential benefits of designing for interdependencies between humans and these new social entities should become fundamental to our design approaches.

When working within a well-defined, human-engineered, safety-critical operational setting, such as a nuclear power plant or an airplane cockpit, exhaustive testing and evaluation can be done. But when the environment cannot be so tightly controlled, as on our streets or sidewalks, it will not be possible to comprehensively evaluate all the types of interactions that may emerge from the introduction of a new robot and their cascading effects. On top of that, the robot not only has to be accepted by everyday consumers, it has to delight them. And at the same time, it has to be observable, predictable, and directable by bystanders with very little, if any, training. Designing for interdependence requires us to shape our design decisions according to, again, what is observable, predictable, and directable; to incorporate the successful elements of user-centered UI/UX design for lightly trained users who expect to be delighted; and to incorporate the systems focus of industrial automation design as well, in order to minimize negative impacts such as safety hazards.

And it bears repeating that the presence of bystanders compounds the challenges. Passersby do not share the same goals as the robot; they start with an impoverished mental model of the robot system; and they must not only learn how the system behaves, but also how to respond to the system effectively—and they have to learn all this with very few interactions that are both intermittent and brief.

THE FIRST STEP IN MAKING GOOD HUMAN-ROBOT TEAMS IS TO RETHINK what design teams themselves should look like. Robot designers can no longer be experts only in the electromechanics of robots and the underlying computer engineering of robot intelligence. Robot teams of hardware and software engineers alone are not adequate. And it isn't enough for them to learn the same UX design principles they have always used.

We saw this in 2018 and 2019 with the decline of Kuri and Jibo, respectively, the first social home robots. Jibo was started by a pioneer of social robotics, Cynthia Breazeal, but even under her direction the product ultimately failed. It focused too much on building an emotional connection with users and not enough on finding ways for Jibo to actually help them. While Jibo advanced the state of the art in human-robot interaction, the company was unable to create a lasting and useful home robot. The negative consequences of focusing too much on the affective design of robots will only worsen when robots become integral to safety-critical activities.

In this new era, robot designers need to incorporate new expertise in the human part of the equation. They need to understand how to support the tasks of their users, how human decision-making changes under different conditions, what biases people have, how they process information in a variety of scenarios, and how people gain trust in others.

And just as the design teams need to change, so, too, does the design process. The traditional walls between AI researchers and roboticists and UX teams will need to come down. They will have to work together from the point of conceptualization in order to scaffold the human-robot partnership from both viewpoints.

In addition, the UX teams need to expand their approach, incorporating appropriate methods from cognitive engineering and human-machine system design that make user understanding one of their top priorities. Rather than viewing the designers of the underlying robotic intelligence and the designers of the user interfaces and "experience" as two opposing processes with a large divide between them, they should be integrated into a seamless process. Design teams should thoughtfully select design methods for a given problem based on the safety-criticality of the tasks, the robot's level of autonomy, the skill set of the users, and time and budget constraints.

Although for most consumer products, user interaction has typically been conceived primarily in terms of the user interface display, in fact user interaction is a critical component of these systems and should be given much more attention. When robot behavior is co-designed with user interaction, there are many more opportunities to design a proper human-robot partnership. The UI alone provides an important set of options to the designer that can facilitate a good human-robot collaboration, but ultimately the number of those options is small with UI alone. For example, UI designers might decide whether a sidewalk delivery robot will use a particular auditory signal or a visual indicator (such as a blinking light) when it approaches a pedestrian. But focusing on the display alone does not allow the designer to make sure the underlying logic of how a robot responds to the presence of a pedestrian is sound. What if the pedestrian does not respond to the auditory or visual indicator? Should the robot proceed anyway, inching forward slowly—that is, attempting to communicate with the robot equivalent of body language—or should it remain at a stop? If it is programmed to simply remain at a stop, it might get stuck numerous times every day; indeed, it might never be able to move, because uncertainty in the movement of the people around it would always be triggering this response. This may be an exaggeration, but it demonstrates the complexity of the design issues involved in trying to achieve the appropriate human-robot partnership. We like to represent the relationship between user interface and the rest of the system design with an iceberg model. The interface is only a small fraction of the system—like the tip of an iceberg—and it happens to sit above the surface because it is so visible to consumers and to the entire product design team. But it's really the robot intelligence and the overall system design that make up most of the iceberg structure beneath the surface.

A process that takes a human-centered approach to the design of robot intelligence opens up many more options. For example, a robot can be programmed to make more distinctions among the types of people it approaches, in order to select an appropriate communication method to use with each one. Perhaps it would communicate with children and adults in different ways, for example. Or it could factor in typical human behavioral patterns when deciding what action to take. This is one way in which even a robot that doesn't understand a lot of social norms could still abide by them. In short, the development of robot intelligence is typically a design activity completed by roboticists, but UX and cognitive engineering experts need to also be included in evaluating decision algorithms, constraints, and user interaction points within the robot's decision-making process.

An even more powerful design process would start with considering the needs for observability, predictability, and directability, both between the robot and users/supervisors and between the robot and bystanders. This means designing specifically for interdependencies as

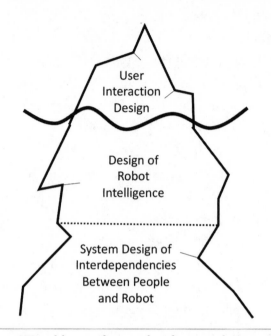

User
Interaction
Design

Design of
Robot
Intelligence

System Design of
Interdependencies
Between People
and Robot

FIGURE 16: UX and human factors often focus on the user interface design and implementation. However, there is limited opportunity to achieve an effective human-robot partnership when the UX design effort is isolated to UI. Human factors should be considered from the beginning as essential to the underlying design of the robot's intelligence, and the interdependence among the robot, the user/supervisor, and bystanders needs to be embedded more deeply into the system.

an intrinsic part of a robot's system. Instead of relying on the UI design, almost as an afterthought, to support the overall performance of the robot system, the entire design would then work together to enable the robot to function collaboratively with humans, so that the objectives of both can be seamlessly achieved. To use the example of a sidewalk delivery robot once again, let's say the hybrid design team has determined that the robot will struggle with ambiguity in the movement of people. The team members agree that there needs to be a way for a user to help the robot get unstuck. There may need to be a tele-operation capability, where an operator can provide assistance from a remote control center when a robot is stuck. From this control center, the operator could monitor a fleet of robots and offer assistance as needed as they canvassed the city. Or perhaps these robots need to be able to determine for themselves when they have become stuck, and signal the control center to ask for help. At this point, the operator could take control and remotely guide the robot out of its confused state.

This is the type of design opportunity that surfaces when you design a system for interdependencies between people and robots from the beginning. Obviously, attention still needs to be paid to the interaction design, and the UI display will support this design option. But you would never discover such a design option by only focusing on the user experience. Similarly, such a process, as in our hypothetical example, may help the team discover new robot intelligence capabilities they would not have otherwise considered (in this case, the capability of detecting when it is stuck and calling upon a remote operator for help). These discoveries could be important to the robot's ability to survive on its own on the streets of the city and crucial to its success.

What does a co-design process look like in practice? Most likely it would take the typical UX phases as a framework but start at the base of the iceberg and work its way up. UI design activities would be triggered as team members identified needs, particularly in the design of interdependencies. The idea is to use an iterative approach in which the designers work high-level concepts and then task design and more and more detailed prototypes, measuring impact and performance throughout the process. The process prioritizes participatory design, an approach in which the designers continually learn from users as they bring the users concepts and prototypes of higher and higher fidelity and develop more fine-grained understandings of the interdependent relationship among the robots, the users, and the bystanders.[18]

The process starts with designing the distribution of tasks across the robot, users/supervisors, and bystanders. In this stage, taking account of how people lose or maintain situational awareness is key. The goal here is to ensure adequate observability, predictability, and directability of the robot's decisions and actions. To achieve this, the product development teams must have integrated knowledge of both the underlying psychology of the users and the full design space of the robot intelligence. In our example, the team would need to know that the robot may struggle in crowded situations, and that too many people moving closely around the robot may cause it to become paralyzed. The UX team alone would not have known to design solutions that address this vulnerability. Likewise, the roboticists may have tried hard to design decision algorithms that could handle these challenging situations, but may not have thought of ways that people surrounding the robot or in a command center could help the robot become unstuck. Only by integrating the two perspectives can a design team forge a truly successful human-robot partnership with interdependencies that work.

From the trend in agile software development, which prioritizes collaborative development team effort and flexibility, we have learned the value of prototyping early in the design and development of a system and testing throughout the ensuing process.[19] This same strategy can be applied here, so that teams can begin to incorporate their ideas for the human-robot partnership early and refine them as they go along. This is similar to Google's design sprint but extends to include the development activities lower in the iceberg model. This is so critical, because it's impossible to predict how people will respond to different hypothesized solutions for complex problems. The only way to get it right is to test the ideas with actual people throughout the design process. Of course, how we evaluate human performance will also have to evolve as robots become more capable of autonomous action.

The iterative "design, prototype, test" cycle is valuable for designing human-robot partnerships, but the methods need to be expanded. Typical UX testing and evaluation focuses on measuring the affective response from users—that is, whether they are delighted—rather than overall performance effects. Also, UX tends to be incremental, producing designs that improve user experience in minute ways over the previous technology, but without subjecting the system to more sweeping safety evaluations or considering more fundamental improvements. This has worked for consumer products that can simply be rebooted when the "blue screen of death" appears. But it's not good enough for working robots. It's not enough to use focus groups to gather qualitative feedback on whether users like a new robot. Human responses must be measured, and it can be done well. Human behavior is predictable and measurable at scale. Human interaction with any product will eventually yield systematic responses and reveal statistically significant trends, but only when there are enough users included in the evaluation. And those trends spring from human psychology and cultural norms. We can design iterative tests of robots to evaluate how the robots will do in a human-robot partnership, and we can use the data we gather to figure out how we need to tweak the design to create a partnership that works well for everyone involved.

Likewise, usability testing is often limited to a small set of scenarios, which are usually benign. However, just as in human-human relationships, in robot-human relationships trust and interdependence are only truly put to the test and validated in times of stress and uncertainty, in events that occur outside of the system under design. Robots of the future must be tested across a wide range of scenarios, including failures or extreme environmental conditions, such as accidents caused by outside entities. Only with a fuller understanding of these stressful situations can we truly provide for the kind of interdependence that works.

So let's put this all together. Below, we'll take a look at how to think about the design process of a working robot from beginning to end. We hope it is quite clear by now that the first step is to understand the users and bystanders with whom the robot will interact.

User Research: We begin with user research to understand the various types of people who are likely to interact with the system, including users/supervisors and bystanders. Users can be interviewed, and preferably also observed, in situations that are relevant to their expected interaction with the system, using methods such as contextual inquiry (a specific process with steps that have been outlined in detail by other researchers[20]). Watching users perform the tasks that you are aiming to automate will surface concerns and inspire insights and solutions that will help guide the design of the human-robot partnership. There are likely many steps that users take when doing tasks manually that are second nature to them. These may include work-arounds, inefficient steps, and innovative approaches to get the job done. People often are not aware of the shortcuts they take, so just asking them about their process may not in fact bring them to the surface. It is only through direct observation and study of how the work is done that these elements can be identified and factored into the human-robot partnership.

We then develop user personas, both for the users/supervisors and for the bystanders. The reason for including bystanders in the personas is to minimize the robot's interference with them as it goes about its job. The design team must try to understand user and bystander goals and constraints, examine the dynamic processes associated with their independent tasks, and hypothesize about their potential interactions with the system. The analysis should involve exploring how human performance will change under varying conditions. For example, different users will have different skill levels, or their workload may vary. What will they do when there are system errors? And this analysis needs to be tailored to specific applications, tasks, and subtasks.

Design of Human-Robot Interdependence: Once the users and bystanders are well understood and modeled, the design of the overall human-robot partnership can begin. The team can start by designing an overall workflow system between the robot and the people with whom it will interact. But this is not just about workflows when everything is going well; it's about how to improve workflow paths at times when things get tricky. This is where the models developed during

INSPECTOR DRONE PERSONA

Primary Goal is to go into indoor, unsafe places and collect imagery of the entire space so that structural integrity analysis can be performed

2 Modes:
1. Fly to a location to collect data
2. Hold position autonomously so the imagery can be collected

Output:
- High-resolution pictures covering all surfaces in the vessel
- Wall thickness
- Measurement and location of all identified defects

Challenges:
- Hot and dirty environment
- Cluttered space with variety of obstacles (e.g., poles)
- Vulnerable to getting lost within vast space

FIGURE 17: An example of a robot persona for a partially autonomous drone that would be used for indoor industrial inspection. This description of the role of the robot will help the entire engineering team remember the key design points.

the contextual inquiry and task analysis phase come into play. The data gleaned from that process can be used to identify breakdowns in how human-only teams execute their tasks. These are the areas where robot assistance is most needed or may be most desirable. Armed with the understanding of the process they've gained from their research, the designers can narrow their focus to developing a robot that can help the humans avoid these breakdowns, lighten human workloads, or fill in troublesome gaps. The resulting models can later be used to evaluate how robot behavior will complement or transform human decision-making and workflow. Indeed, in later iterations, other kinds of breakdowns will come under scrutiny, including breakdowns in robot understanding, in robot operation, or in human understanding of the robot system.

To facilitate the initial design of the robot's behavior, as part of this process we develop a robot persona. We know that people personify the robots they interact with; therefore, we must explicitly design robot personas to set the vision and design the robot to calibrate our human expectations. User personas are developed to help a design team empathize with a user population. Robot personas can provide a vision for the design team for the other half of the human-robot partnership equation. This persona will enable the roboticist to keep the human-robot partnership at the forefront as the robot software is developed.

The design team should use the Situation Awareness–based Agent Transparency model and the three requirements for interdependence (observability, predictability, and directability) to design the robot persona and model the robot's tasks. All of this sets the stage for an effective robot-centered design effort that results in an effective human-robot partnership.

Decide, Prototype, and Test: The previous two steps lay the foundation for the iterative design and test cycle. Here, classic UI prototyping and testing methods need to be expanded to prioritize robot-human interdependencies. The various options for robot behavior stem from that overall framework of interdependence. And these design approaches must be tested across a wider range of scenarios than is typical in today's design processes. These scenarios need to include failures, or extreme environmental situations such as accidents. Some of these accidents may be caused by outside entities. The goal is to maintain optimal human-robot interdependence in these critical moments. In times of stress or anomalies, the human's trust in the robot may flag, just as in human-human relationships. Interdependencies may be put to the test. Therefore, it is at these moments when it must be clear to the human what the robot is able to do and how it will do it, and what the human must do to bring about the best result.

There is a rich array of UX and human factors approaches for designers to draw from at this stage. Below we summarize two categories of the methods currently being used across consumer products and industrial systems.

1. **Expert analysis to predict a user's response to a design concept**—Experts in this approach evaluate the ability of the robot system to support the needs of the user. This method involves the use of heuristic design principles to guide the design of various features of the system, such as the right level and timing of feedback, and efficient navigation through menus. Expert evaluators ask questions that they then attempt to answer, such as "Will the user try to achieve the right outcome?"[21] They may use hazards analysis techniques to determine how the system may fail, and to ensure that designers mitigate these failures by designing solutions into the workflow patterns of either the robot or the user. Another approach is called GOMS (which stands for Goals, Operators, Methods, and Selection). GOMS aims to model key aspects of human information processing to analyze, for example, how long a person may take to make a decision.[22] Finally, a family of techniques for cognitive modeling aims to simulate how human cognitive processes will impact the system. These techniques may involve prototypes of solutions and tasks with differing real-world fidelity.[23]

2. **Observation of user behavior when using the prototype**—It is impossible to perfectly predict how a user will respond to any given feature in the human-robot partnership. So there are many ways to test people's responses and then factor the findings from this testing back in, to refine the design. These methods often involve recording user behavior and performance during interactions with a prototype or with actual implementation of a system. Data may be gathered from a range of interactions. Users who are participating in the test may be brought into a laboratory to try out different options, or data may be collected about usage during live deployment of a system (beta testing). When users are brought into the lab to try out a prototype, they may be asked to give feedback, or to "think aloud" while performing the

required tasks. In the "think aloud" version, they speak aloud to explain what they're thinking and doing during their interaction with the system, and this is usually recorded for later analysis.[24] Once a system goes live, evaluation can continue. Tools similar to Google Analytics can be used to track usage of the system and any problems users are having. Such tools can provide great insight for future features or software upgrades.[25]

The cost of these evaluation options can vary significantly. While some of the more costly evaluation approaches are appropriate for the design of complex industrial applications, they may not be practical for all consumer systems. However, that is not an excuse for forgoing evaluations of the human-robot partnership as it is being developed. Given the range of alternatives and their applicability to early designs, prototypes, and fielded systems, developers should be able to select an approach appropriate for their product. Indeed, not engaging in evaluation can also be costly, because it is likely to lead to problems when the system is fielded.

We had an opportunity to apply this unified approach to designing a new drone for inspecting hazardous indoor environments. Other drone developers were applying the classic approach, of having one team design the drone's capabilities and another team design the user interface for controlling the drone. Applying our hybrid design approach led to a substantially different design for robot intelligence and human-robot interaction than we would have pursued otherwise. The new design provided more help to the user than the designs that came out of the classic approach.

For example, we found that what users were most interested in was capturing data about the indoor environment, so they could determine whether there were any compromises to the integrity of the structure. They were not interested in flying the drones; they just wanted the data. Implementing a fully autonomous drone was still out of reach for the industry, however. So we came up with an alternate approach that gave the users what they wanted while filling in the gaps in robot capability. Users wanted the system to operate like a camera on a pole—and we knew we could design a system where the drone could hold a single position for some period of time without any human control. We dubbed this mode the "autonomous station-keeping mode" and designed a workflow that allowed the user to shift between flying the drone (to navigate the indoor environment and find a specific location) and holding a location (in this new station-keeping mode). This design allowed users to take their hands off the drone controls completely when gathering data and simply pan and zoom the camera to take detailed pictures. Instead of trying to achieve either fully autonomous flight or full remote tele-operation, we had come up with a new approach.

In doing so, we avoided introducing yet another clunky robot into our world that no one would quite know how to use. Instead, we created robots that partnered seamlessly with the user, able to navigate all the challenges the world would throw at them.

CHAPTER SIX

How Do You Say "Excuse Me" to a Robot?

H OW DO YOU CONVINCE A ROBOT TO GIVE YOU WHAT YOU want? Countless books have been written on the art of persuading people. But people have free will. Why should we need to persuade robots? Won't they just do what we want them to do?

New intelligent robots learn and adapt, much like people. Their behavior is constantly changing. This is a good thing, since smarter robots can do more for us. But it is also a lot harder to predict how to influence the behavior of a smart robot than a simple one. Just as with people, in order to make sense of what a robot is doing in any particular moment—and exert some control over it—we need to understand how the robot thinks and why it behaves the way it does in general. The design of the robot will determine whether we can gain that foundational understanding. So the question is, how do we design robots that people can intuitively understand and influence?

Every object, no matter how sophisticated, has to have ways of letting users know how to use it. These are the *affordances* mentioned earlier, and the clearer the mapping between what the user notices about the object and how they can use it, the better. Historically, affordances were thought of as characteristics in the appearance of a device that provide clues to its function or use—such as a door handle to help a user know how to open a door, or a large metal plate on the side opposite the hinges letting you know where to push, or a large vertical handle letting you know where to pull. As digital systems came into our everyday lives, affordances were expanded to include other physical characteristics, such as the shape of buttons, the placement of switches, and digital displays—such as icons and menus. The volume up and down buttons on a remote control are in the shape of up and down arrows for a reason: it makes it clear which arrow increases the volume and which one decreases it.

Robots have affordances, too. But designing appropriate affordances for automation is not as simple as selecting and displaying a static set of options for interacting with the system. Much like a person, a collaborative robot is an independent agent that changes the way it responds to user input based on the situation. An instruction in one situation may be different from the same instruction in another situation. Even a robot as simple as the Roomba exhibits this principle, because pushing the big round button has a different effect depending on whether the Roomba is docked, stuck on a rug, or low on battery. Automation affordances have to present more options than other types of devices. We don't just interact with them in a single type of situation. And the affordances go beyond just its physical form, because there is now a more complex user interface, or even, possibly, voice or gesture controls. The designers must make all of this transparent to users—as well as bystanders in many cases—so that the humans interacting with a robot know what to expect of it across a number of possible situations. The affordances must give us clues about how to interact with the system. That is, they must somehow shape our mental model of how the robot thinks, how it behaves, and how it responds to human interaction. And all that has to happen before you even have a chance to directly observe the robot's behavior in a given situation. Bystanders will have to know how to interact with a type of robot they've never seen before, and they'll have to know how to do it intuitively and quickly.

So much of what we understand of how other people think and behave is implicit—we form our understanding and judgments without even a conscious thought. For example, say you're at the grocery store, and as you unload your groceries onto the belt at the checkout, you can't help but notice the man paying in front of you. He has a broken leg and is on crutches. While taking out his wallet, he drops his car keys. You will most likely stop unloading your own groceries and pick them up for him without uttering a single word, except for a quick "You're welcome!" after he thanks you. The full leg cast and crutches are a type of affordance that quickly clues you in to many things about his limitations, making it clear that he can't bend over to pick up the dropped keys. There is no need for him to explicitly ask for help.

As mentioned previously, in industrial applications, operators receive hundreds, or even thousands, of hours of training to learn about an automated system before they actually start to use it, but the average person on the street won't have that luxury before using the many kinds of robots that will start showing up in our lives. To overcome this challenge, designers will need to design affordances, and making them simple and intuitive will take some work. One way to make them intuitive is to consider the norms that inform how we interact with people, and figure out what these norms may be able to offer us in designing working robots. But which of our existing mental models are most appropriate for intelligent robots? And what mental models will people have to learn from scratch?

Our research has identified two types of situations that will require different types of automation affordances. The first is when time is of the essence, and quick action by the human or robot is required to avoid or respond to a safety hazard. In these cases, automation affordances will need to be instinctive to the human user or bystander. The second type is when the person has time and motivation to interact with the robot to troubleshoot and problem-solve together. In these cases, the affordances must support more deliberative interactions.

SPEAKING THE SAME LANGUAGE

If men are from Mars and women are from Venus, then robots today are from a black hole. They think differently than humans, behave differently than humans—and it can require mental gymnastics to figure out how to get them to do what you want them to do. The decisions of some of today's most advanced artificial intelligence systems remain a mystery to the very designers of those systems. Machine learning allows a robot to develop its own internal mathematical representations to relate what it sees as inputs to what it produces as outputs. To make sense of the world, we use commonsense knowledge about cause and effect. Many modern AI systems do not do this: they boil everything down to correlation.[1] They simply map input to output, using what we call machine learning, but there are many different ways for them to do this mapping. Only some of these possible mappings capture causal relationships that we understand and use to make sense of our world. The problem is that it's hard for a person to inspect the output of a machine-learning system and figure out whether the machine is learning a spurious correlation or has uncovered a new important relationship. The problem is compounded because many of the most widely used machine-learning models—including artificial neural networks—do not make sense to even highly trained developers.[2] As a consequence, while a developer may be able to demonstrate that the system's performance in, say, classifying situations or objects, or predicting outcomes, is acceptable for a particular task, she does not know precisely why it makes each of the decisions it makes, or whether the AI's "reasoning" corresponds in any meaningful way with a human's reasoning about the same problem. Even with simple, rule-based automated systems, where the logic of the system is completely determined by programmers, we often struggle to understand how to interact with the system to achieve our goals. Think about how challenging it is to program your washing machine to start later in the day, or how many steps it takes to set up a home speaker system that is controlled from your phone.

By and large we make things work, sort of, as long as we have the luxury of time to learn through trial and error what a device can and can't do. But what about those moments when what you need is for a robot to follow your directions immediately, when you don't have time for the mental gymnastics to determine how to translate your request into robot language?

Typically, when humans need to interact with robotic systems, there is in fact little time to engage and give commands. So in many situations an affordance needs to support a quick reaction and redirection of the robot's behavior. When a robot is buzzing down the hospital hallway to deliver a medication for an urgent situation, and a nurse pushing a stretcher with a patient turns the corner, the nurse needs a way to make the robot stop quickly—for example, by yelling "Stop!"—before it crashes into the sick patient. During these safety-critical moments, users and bystanders will need to communicate in ways that come naturally.

When something is about to crash into your patient, yelling "Stop!" is an instinctive response. The situation triggers a cognitive process, and that same cognitive process should work to pick the right response to the robot. In other words, the right command to the robot should be the same one the nurse would naturally use for anything else heading pell-mell toward the stretcher. The robot's automation affordances must be designed to be in harmony with what the psychologist Daniel Kahneman has called "fast thinking."[3] Fast thinking is what comes most naturally to us in these kinds of emergency situations, and it is an instinctual and unconscious response. You are engaging in fast thinking when you swerve as a car unexpectedly enters your lane on the freeway, or when you turn your head when your name is called in a crowd, even though you weren't aware of listening. It is automatic. Kahneman contrasts it with more analytical, logical "slow thinking." Our brains use fast thinking in emergencies because they are hardwired to respond quickly to such situations. They match patterns in the environment with an effective action without conscious thought. At these moments, we don't have the time to go through a slow process of recalling from memory the various ways we can communicate with the specific robot, or to press a bunch of buttons on a control panel. We need to be able to respond in a way that will affect the robot as soon as we see the potential for danger. For these tasks, the onus is on the robot to understand and respond to human language.

To a large extent, we have already designed affordances for the things we use in our everyday lives that conform to the social norms of interacting with people. When a car crosses your path and it's too close to you, you honk your horn. When bicyclists approach people from behind on the bike path, they ring bells on their handlebars. If you see someone about to cross the street, and he doesn't seem to notice a car approaching, you yell "Watch out!" Our already well-honed communication strategies enable us to engage in fast thinking and have a quick reaction.

Robots of the future will need to know what a horn means, and what a bicyclist's bell means. They will have to understand our verbal and nonverbal human language. Asking humans to uproot decades, or even centuries, of ingrained norms is not practical. And robots will need to know all of this within the context of countless situations. For example, if a horn honks from a car in front of you, you respond differently than if the honk comes from behind you. If you hear "Watch out!," you immediately stop, and then look around to determine whether you're the one who was being warned. If someone yells "Stop" while you are taking one step off of the sidewalk to cross the street, you may step backward onto the sidewalk instead of even completing your step.

Our intuition helps us connect those warning sounds and words into fail-safe behavior without conscious thought. Robots will need to be designed to work with that same intuitive, fast-thinking model. They will need to know how to give us the same kinds of cues that people give to other people, and they will need to be able to follow the same kinds of fail-safe actions in response to us—or better ones. For robots to truly become social entities in our world, they will need to behave in predictable ways so that we know how to influence their behavior.

What happens if we don't get this right? It can be hard to imagine what would happen if we lost the ability to predict how others behave. We all work so hard to build a world that feels predictable to us. But have you ever entered a butterfly garden with a toddler? It is just impossible to predict where the butterflies will flutter. Adults are delighted, not disturbed, because we know that the butterflies are harmless and where they fly is of no consequence to us. But some toddlers are completely and utterly terrified. The butterflies are unpredictable in their movement, but the child may not know that they are safe. After all, other flying things, like bees, are dangerous. And the butterflies are everywhere you look. The terror that the toddler feels in the butterfly garden is a preview of the anxiety we would feel living in a world with robots when we have no easy way to understand and influence their complex behaviors.

Of course, it would be impractical to conduct studies on every form of human interaction in every possible situation that may arise, in order to design an encyclopedia of appropriate robot responses. The robot's behavior, just like human behavior, becomes impossibly complex to codify manually. But there are two emerging ways we can get around the complexity of human behavior and robot behavior. First, we can have robotic systems learn as much as we can teach them about human behavior at scale with machine learning, and second, we can design dynamic automation affordances by enabling robots to think and behave more like people do. Neither approach will be sufficient on its own, but together they give us a fast track to designing robots that behave reasonably.

In the first approach, we would apply machine-learning techniques to actually learn human behavior and use this to program robots to behave in a similar way. There is already some evidence that it's possible to train neural networks to conform to and predict human behavior.[4] This includes predicting all sorts of behavior, ranging from whether a driver will turn at an intersection to whether two people will hug or shake hands based on their body language toward each other. The tricky part is that effective collaboration with human partners requires more than just predicting their motions or movements. We are only able to think ahead about what they might do or need because we simulate in our own minds what their goals and preferences are, how they are feeling, and so on, inferring mental states that we cannot directly observe. A machine cannot learn to simulate these internal mental states of ours without some help and guidance from us. Our lab at MIT has been working to crack this problem by developing AI that can learn to infer human mental states and implicit norms of behavior through observation along with a query-and-response dialogue with a human partner.[5] We recently conducted experiments where we had one of our intelligent robots work with people to make sandwiches together. We demonstrated that a robot using our AI to infer a human's mental state was a substantially better collaborator than a robot that did not model the human. We still have a long way to go to learn how to do this at scale, and figure out how to apply it to develop robot decision algorithms on embedded computational platforms. But laboratory research shows that this is a possibility in the not-too-distant future.

Another way to ensure that a person can instinctively understand a robot in critical moments is to design affordances that leverage an existing, dominant mental model. For example, when the personal computer first came out, the designers adopted a model for the interface that we are all familiar with. They organized data on our computer like files in a folder and made searching for those files as similar as possible to the physical search through a file drawer.

We need a similarly easily understood model for robots in time-critical and safety-critical moments. But what is the right analogy for robots? Although robots are not like file cabinets, there are many other possibilities. A dominant theme in culture, however, is that people naturally personify robots.[6] When a robot shows up in a science-fiction TV show or movie, we expect it to act like a person and interact socially like a person (when it's not trying to take over the world). We propose to build on this natural mental model—not necessarily by building robots that look more like people, but by building robots that think more like people. We view robots in ways that are similar to how we view other people, giving them names, referring to them as "he" or "she," even talking to them when we know they can't hear us or respond. And while we don't want to overindulge this tendency, for reasons we covered in earlier chapters, it's still unlikely we will ever fully break the habit. In fact, research shows that people treat robots more like other people than like machines. This notion was formalized in 1996 in what's referred to as "the media equation," which describes how people tend to treat nonhuman media (robots, TVs, computers, and so on) as if they were people. Through the results of numerous psychological studies, the authors of the 1996 publication, *The Media Equation*, showed that human interaction with nonhuman media was similar to that found in real social relationships. For example, people were polite to machines, even when they claimed they were not intentionally being nice to them, and people treated machines differently depending on whether the machine used a female or a male voice.[7] The authors showed that when people were conditioned to view a machine as a teammate, they were more likely to view the machine as being like themselves and to be more cooperative with the machine.

It turns out that this tendency can be used in powerful ways in the design of automation affordances, because we have a wealth of experience interacting with other people to draw on. Every person is unique, yet, despite our differences, we are still able to quickly figure out how to interact with strangers. When a stranger looks lost, we point her in the right direction. The reason we have this capacity with people is that, despite our differences, our brains are hardwired to think and behave in largely similar ways. Some of this varies, of course, by culture. But as the psychologist Gary Klein has observed, once we have expertise on a given task, we develop an intuition that allows us to match the patterns we see in a new situation with similar patterns we've seen in the past— and we do this subconsciously. We don't have to have seen the exact situation before: we can match the new situation to similar past experiences and fill in the gaps, or fix up what worked in the past for the new situation. This model for cognitive process, called *recognition-primed decision-making*, is especially useful in characterizing complex and nuanced human decision-making in situations where a person needs to make a decision quickly, with only partial information, and the goals are poorly defined. The model has been validated with nurses, firefighters, chess players, and stock market traders.[8]

We can leverage this basic model of human cognition to design robot affordances for instinctive interactions. Robot designers can harness patterns that people are used to seeing in their world beyond the robot, and then design affordances in how they interact with the robot that loosely couple with these patterns. Again, people can tolerate slight variations in the patterns, but if they are completely decoupled from the patterns we are already conditioned to understand, we will struggle with understanding the robot.

These are not the traditional *static affordances* we are used to seeing—that is, the straightforward mappings of affordance to behavior that exist in the kinds of objects we use every day. Instead, by designing robots that behave as we do, we will provide *dynamic affordances*, which will allow us to understand and influence robots and direct their behavior. And we will be able to do this because we will be able to mentally simulate what the robot will do. And we will be able to continually update our model of the robot behavior as the situation progresses, and intuitively know when and how to influence it in a problematic situation. We have an example of dynamic affordances already in how pedestrians predict the actions of another driver. There are static affordances that could be used, like flashing your lights or waving the pedestrian across. But studies have shown that pedestrians primarily use the kinematics of the vehicle to make their crossing decision, because they form models of driver intent by observing vehicle kinematics.[9] Simply put, the vehicle's behavior enables pedestrians to form a mental model of the driver and predict the best way to interact with the vehicle at a crosswalk.

This approach can be translated to robots. Research supports the conclusion that robots that follow social norms, even in part, are perceived to be more intelligent and understanding than robots that are not programmed to do so, and that people are more comfortable and willing to interact with them.[10] For example, our lab has studied for many years how to make robots smart enough to safely and efficiently work alongside human workers in a manufacturing plant. In one series of studies in which such a partnership occurred, we found that people were better able to "get in the groove" of working with a robot when they had first practiced the task with another person, and then worked with a robot that moved and worked in ways that were very similar to how the human partner had moved and worked.[11] Our study participants became very disoriented when they had to work with a robot that moved in different ways each time it completed a task (even if the robot was always following the most efficient path), so much so that the negative experience colored all their interaction with the robot from then on.

We ultimately had to redesign our study to make the robot's movements more predictable. We will need to design our working robots with the same considerations as they maneuver around us on our streets, sidewalks, and shop aisles.

This analogy is critical for more than having people simply accept the robots that surround them in their everyday environment. Helping users/supervisors and bystanders predict how they can persuade and influence a robot's decision-making and actions is crucial to the success of the robot in carrying out its duties as well as in helping us accomplish tasks. Just as our instincts allow us to know how to influence the actions of other people, we'll need to know how to intuitively influence the actions of robots if they are going to be partnering with us at the office, at hospitals, at stores, in the military, and so on.

For robots to truly become socially aware entities that we can influence, they need to behave in more predictable ways. Roboticists typically design robot decision-making to achieve the very best decisions they can and to take the best possible actions given the information they have. However, sometimes a less optimal path or a different sequence of actions may be better at mimicking how a human would approach the problem. If so, that option is the better choice, because it will prevent confusion in the robot's human partners. And again, this is the type of approach that can only be brought out with a hybrid design process that explicitly designs interdependencies into the system.

Designing an appropriate dynamic automation affordance—one that allows us to mentally simulate what the robot will do under a wide variety of situations—is not a straightforward task. Robots think and learn in a fundamentally different way than people do. They see the world differently from people—through different sensors—and record the world with a different form of memory and recall system. Even if they were to "live" the same experiences, the manner in which robots and people drew from those experiences to make decisions and act would be different. Robots are programmed, and we have evolved. And yet we need to design an instinctive way for us to imagine what a robot might do in a new situation.

One thing we know about people is that they think with examples. We match our current situation to our closest previous experiences and make a decision about what to do based on what worked or did not work well in the past. If you are cooking with leftovers from the week, you might look at the ingredients on your counter and try to think of the last recipe you made with similar items. Experts solving some of human-kind's most important decision-making problems, such as controlling a forest fire, use the same strategy. A captain in charge of a firefighting crew will use recognition-primed decision-making to figure out a strategy for a forest fire, for example. Knowing the particular circumstances of the current forest fire, he compares them to his most similar previous experiences, where certain strategies for containing the fire were successful but others were not. This allows him to quickly decide how to allocate resources to get the fire under control, without needing days or weeks to analyze all the possible approaches.[12]

In our lab at MIT, we asked whether a person might more instinctively understand and predict the decisions of a machine if the machine "thought" with examples too.[13] To test the idea, we taught a machine to cluster and categorize cooking recipe data sets. In one setting, we used a standard state-of-the-art machine-learning algorithm to cluster the recipes, which considered all the recipe's ingredients in an unordered list. We then designed an alternative machine-learning algorithm that enabled the machine to define its categories based on one quintessential recipe of the grouping. For example, the algorithm grouped numerous chili recipes together and explained the new category in terms of a particular selected recipe named "Generic Chili Recipe." The system explained that this particular recipe was quintessentially of the category because it included the subset of the ingredients beer, chili powder, and tomato. With the standard machine-learning algorithm, the system tried to explain its groupings with subsets of ingredients too, but the ingredients might have been drawn from across a few different recipes, making it harder for a person to quickly imagine the dish that corresponded to the ingredients.

We then asked a few dozen people to come into the lab and gave them a set of recipes. We asked them to select the category that a new recipe belonged in from the groupings generated by the machine. People answered much more quickly and much more accurately when the machine explained its selected groupings using ingredients from example recipes, as compared to using the standard machine-learning approach (86 percent accuracy versus 71 percent accuracy, a 15 percent improvement). When the machine "thought" using examples, in other words, the participants could use their innate recognition-primed decision-making strategies to predict the machine's decisions.

Redesigning a robot's decision-making process to emulate our own thought processes, such as recognition-primed decision-making, is one type of dynamic automation affordance that can make it easier for us to project how a robot will behave in a new situation.

REVERSE-ENGINEERING THE ROBOT'S MIND

Of course, not all interaction is based on split-second instinctive communication and response. Sometimes we take our time to get to know one another, ask each other questions about our pasts, about what we plan to do in the future, or about what we are thinking in the present. If we are puzzled by a teammate's behavior, we might address the issue with her directly, in order to come to an understanding of how we will do things together in the future. If we are trying to solve a problem with someone and he does something unexpected, we may ask him why he did that, or about what he plans to do next, so we can tailor our own actions accordingly.

Similarly, we will need to have the ability to "converse" with intelligent robots to understand them better or fix a problem. With industrial applications, there are many ways for a user to troubleshoot automation, such as by following detailed diagnostic procedures, or soliciting help from an expert. Tools are provided to help a user choose from a set of possible actions—for example, through checklists, prompts with step-by-step instructions, or a help menu with contact information for technical support. Using these options sometimes—but not always—requires extensive training.

We can't expect everyday people to read an extensive user's manual or take a college course on machine learning to prepare themselves for the coming of street robots. But we can make use of a different type of automation affordance to help, one that supports what Daniel Kahneman called "slow thinking,"[14] the effortful, calculating, logical way of thinking we use when faced with analytical problems, such as calculating how much of a tip to leave on a bill, or figuring out how to maneuver into a tight parallel parking spot. For situations that require a user to analyze a sequence of actions by a robot, or debug a problem, the user will need to be able to draw on their slow-thinking skills to figure out the robot's thought process. Indeed, robots of the future will need to be able to take questions from people about their logic and code, and explain to users and bystanders, in a comprehensible way, what they understand about the world, what they plan to do, and how we can interact to change their plans and behavior.

Our work at MIT enables an autonomous agent to do this translation. The agent we developed can explain a machine's code and logic to a human in an understandable way to give a description of its behavior.[15] We developed a series of algorithms that enables the agent to answer behavior-related questions, to concisely describe its previous experience testing its behavior under various environmental conditions, so that the person can determine under which conditions the behavior does or does not occur (for example, "When do you do __?"; "What do you do when __?"; or "Why didn't you do __?"). In doing so, we provided people with the capability to develop and refine their expectations concerning the behaviors of otherwise opaque autonomous agents through direct inquiry. This interactive, targeted expectation calibration is an important initial step toward understanding and being able to debug robot behaviors, as it facilitates precise identification of the ways in which expected and realized actions do not align.

WHEN OTHER MODELS BREAK DOWN

We've described the opportunity to build upon human-human interaction models in terms of communication, behavior, and even thinking as well as other models that we are familiar with, such as the desktop model used for PCs, to improve our interaction with automated systems. But what about when there are new interaction needs that require new models?

We saw this with touch screens and mobile devices. The early Palm Pilot, for example, included a stylus, to mimic the way we use a pen for paper. But designers soon realized that this wasn't the best solution, because switching between the physical and virtual world created hurdles—such as the additional time it took to take out and return the stylus. Instead, designers went to work designing new gestures for direct manipulation that eventually were adopted across the industry. A great deal of user research went into the design of direct manipulation gestural interaction to make it seem as natural as possible. In fact, Apple is widely known to invest significant funding in user research and user interaction design for their products. So it's not a surprise that we've seen family members with limited computer or technical experience being able to pick up and use iPads or iPhones with incredible ease, even without training on how to use them. Moreover, the entire industry has now adopted these new interaction methods. Intuition works in the use of these devices regardless of who the manufacturer is. On our phones and tablets, we now pinch our fingers across the screen to zoom in and out without even thinking about it.

Robots have fundamentally different capabilities than people—this is one of the strengths of working with them. Which means that the human-interaction analogy can only take us so far. And there won't always be a convenient analogy for the interaction, such as in the file-cabinet analogy for desktop computers. We will need to design and engineer new affordances, just as the mobile computer industry did when moving from the stylus to direct manipulation gestures (e.g., pinching and swiping across the screen).

The automotive industry knows a thing or two about standardization. We can rent a car we've never driven in a new city and safely and reliably drive it out of the lot and to our destination. We feel a little rusty at first. It takes us longer than usual to adjust the seat and mirrors and figure out how to turn on the blinkers and windshield wipers. But we are soon on our way, as if we'd driven the car for years. This kind of seamless transfer is only achievable through standardization of the controls and interfaces. The controls that are needed for split-second decisions, such as the steering wheel and brake pedal, work the same way in all cars. The subtle differences are only in the controls that aren't involved in split-second decisions, such as the wipers, blinkers, and lights. But even for those, the differences are minor. The headlight switch may be in slightly different place, but it's always in the same general area somewhere around the steering and dashboard, and there are always options for low and high beam. The variability across different manufacturers is cognitively manageable.

The controls and displays in our cars are heavily influenced by the Federal Motor Vehicle Safety Standards, and having to take a driving test to get a license ensures at least a basic level of training for drivers nationwide. Will the robots of the future have the same level of standardization for their controls and interfaces? Will robot users have to pass a robot operation test to prove that they have learned new robot-specific affordances? There would be proven benefits for both operation and safety issues, but there would be a cost. Moreover, bystanders still wouldn't get the training—after all, they're not exactly given a choice about whether they want to share the sidewalk with robots, at least not in the same way that people choose to get driver's licenses. And robot technology is advancing so rapidly that training protocols would struggle to keep pace.

Still, in some cases, perhaps this idea is a good one. Take flying robots, for example. Numerous companies are developing, testing, and fielding systems for delivering packages to you right at your door. The people who load the package onto the drone and send it on its way surely should have some training for the job. And people who fly recreational drones could be required to pass a test proving that they know to avoid interfering with safety-critical happenings on the ground, such as when emergency vehicles have been called to the scene of a disaster or crime.

But if you have simply ordered an item and the drone is hovering toward your front door, how will you communicate and interact with it to tell it where to place the package, or to warn it to stay away from your dog or child? What about bystanders, who just want to make their way around it as it hovers ominously in front of them? We have no clear human analogy for interacting with a free-flying robot. Even human-animal analogies break down. When was the last time you tried to gesture to direct a bird flying toward your bird feeder? We need new models for affordances for these systems. The tech industry is working on this. Amazon filed a patent in 2016 for a delivery drone that responds to a person waving their arms at it. But it will require iteration to get these new affordances right.

The design problem is a tricky one. How can the robot quickly tell people where it intends to fly or what it plans to do next? Many of the "copter-like" systems with rotors can fly in any direction. It's not even clear which part is the front or back. People may feel like it's going to crash into them—in fact, if there isn't a good way for them to influence its behavior, they could be right about that.

Robotics researchers are actively investigating the best ways for robots to communicate their flight-path intentions to nearby viewers. Interestingly, one of the best strategies so far is a bit counterintuitive. It's to make the flying robot *less capable*, but more predictable. The idea is that you can simply constrain its flight behaviors, using biological and airplane flight as inspiration. In this way, a drone flying through space looks more similar to other systems we see flying, and triggers a mental model that helps us understand the new system's behavior.

Similarly, drones need to have new signaling mechanisms to visually communicate directionality, designed in conjunction with the flight behaviors. Studies have shown that this type of joint design can work, enabling people unfamiliar with a system to predict a robot's intent, but that different types of signals have different benefits and drawbacks, with trade-off in terms of the precision of the person's prediction, generalizability across robot platforms, and perceived robot usability.[16]

Automation affordances need to go beyond static mappings to incorporate dynamic affordances. It is only by taking account of the dynamic nature of robot movements that we can ensure clear mappings of a robot's imminent behaviors.

Because robots have agency, and can learn and adapt, the human-human interaction model should still be the basis for automation affordances. Robots should be carefully designed to be consistent with human decision-making and behavior, and they should even be programmed to think something like people so that a user can better predict what the robot may do in a wide range of situations and inputs. We can divide the types of situations into two types, based on the fast- and slow-thinking models:

- **Instinctive situations:** Robots will need to mimic human intuition to receive brief, natural human communication and respond with fail-safe actions without lengthy interactions. To ease the mental burden for people of interacting with robots, we will need to design robots to behave in ways that people can intuitively predict.

- **Deliberative situations:** Robots must be able to explain their actions and rationale to the user through direct inquiry.

When humans interact with robots, we enter new territory—we're going beyond human-human interaction into a new world where one of the parties is not in fact human. So new interaction models need to be carefully designed. We can draw on human-human interaction whenever possible, but sometimes we'll have to draw on other models. And the best way to make sure working robots remain reliable and easy to work with is to standardize these models as they are developed and proven out.

CHAPTER SEVEN

Robots Talking Among Themselves

MOST OF THIS BOOK IS ABOUT HOW TO DESIGN ROBOTS THAT can behave well enough to work with us. But there will be some tasks that robots may just be better at than us. It's equally important to design robots to *not* work with us, when appropriate.

As we've discussed, there are many situations involving a working robot where you simply cannot detect a problem and react fast enough to prevent a catastrophic failure. There are also situations where trying to include a human in the decision-making loop or negotiation is not ideal, because the robot is just plain better at doing the task than a person. In these cases, the decision about who should be in charge should be simple—surely the robot should be designed to detect and address the situation with minimal or no human input. But even in these simple, seemingly obvious situations, we often get the design wrong. That's because we're forgetting something important about human-robot collaboration.

Take, for example, a deadly airliner crash in 2002. One evening in July, a commercial airliner was flying from Russia to Barcelona with sixty-nine passengers.[1] It had reached a cruising altitude and was just crossing over the German border when it began approaching the path of another jet delivering cargo from Italy to Belgium. The two planes were on a collision course, and the air traffic controller was busy with other tasks needed to safely manage the airspace. He didn't notice the impending collision until less than a minute before impact. At that time, in a panic, he made a call to the pilots: "Descend" to one, "Climb" to the other.

Both aircraft also had automation systems on board that could detect such conflicts and coordinate directly with each other to determine a resolution, without any input from the pilots or air traffic controllers. This Traffic Collision and Avoidance System (TCAS) was designed for exactly this scenario.[2] Both of the pilots simply had to accept the computer's recommendation.

That night, seconds after the air traffic controller gave the pilots commands, TCAS gave the pilots the opposite commands: "Climb" and "Descend." One followed the automation instructions, and the other followed the air traffic controller's. They descended right into each other. All the people on board both airplanes lost their lives that day.

This incident changed how commercial pilots are trained to follow commands while airborne. Now, they are instructed to only follow the commands of the aircraft and not those of the air traffic controller—the outsider. That's because people cannot always be relied upon for timely negotiation between robots in crisis situations and because, on average, robots keep a leveler head than humans.

Problem solved. We learned our lesson from aviation. But oddly, a similar issue now crops up with autonomous cars. Recently two driverless cars came to a stop at the same intersection, both of them waiting for the other to proceed. Because there was no means of direct communication between the cars, and because the backup drivers on board had no way to make eye contact, motion the other forward, or signal each other, the two autonomous cars were paralyzed, stuck in a deadlock.

These two cars were made by the same manufacturer, and the designers could potentially have designed them to communicate directly with one another. But they didn't. In commercial aviation, the root of the problem was a conflict between the human's directions and the machine's, and we learned the hard way that the human couldn't be trusted. But there was no conflict in directions with the autonomous cars; they were simply left directionless. The problem was that the designers thought the cars really only needed to conservatively yield to other human-driven cars. It didn't occur to them that the autonomous cars might come face-to-face in this way, and might need to be able to talk to other autonomous cars.

The future is even nightmarishly more complex than that. Imagine robots of different types, from different manufacturers, buzzing down our sidewalks and streets, flying overhead, all speaking different languages internally, with no ability to coordinate or communicate with each other. They won't just be in our space—they'll be in each other's space as well. What then? Every time they come face-to-face with each other or get stuck, will they need one of us to intervene?

Say that a truck that serves as a hub for delivery robots pulls into your neighborhood. The driver gets to your block, climbs into the back of the truck (which is also a command center), and begins dispatching the robots and drones. Rather than acting on their own, the robots help each other out. One finds a snow heap or fallen branch in the way and shares its location with the others, so they can plan routes around it. They coordinate their paths to avoid collisions and to ensure there are not too many robots on the same block at the same time, because the designers didn't want them to become too much of a burden to the people in a neighborhood. When one robot or drone is unable to get around an obstacle or climb stairs, another one comes to assist—it can move the obstacle or nudge the robot over the crack in the sidewalk. A larger robot zooms down the street to get packages to driveways, then a smaller one, scaled to the size of the driveway or walkway, takes over and navigates the tighter space more slowly. A drone lifts a package off of a robot and carries it over surfaces that are hard to navigate. If other robots are in the area, the delivery drones might ask them for help, or try to minimize their interference. For the most part, they simply communicate with each other. The supervisor, meanwhile, stationed in the command center, is more like a conductor, directing the performance of the team of robots and drones while they work in harmony. This is the picture of everything working just right. As depicted in figure 18, the robots share information about their understanding of the world, and the heterogeneous fleet and the supervisor work together seamlessly. They manage to do their work without burdening others around them, just as a human delivery driver does when making rounds. In reality, today we can barely coordinate two autonomous cars in a standard intersection.

The question is not simply, Do robots need to be able to directly communicate with each other? It's how they should collaborate with one another and when, what they should communicate, and what our role is going to be once robots are dealing with each other directly.

Designing for clear and effective communication between working robots is as important and complex an issue as designing for clear and effective communication between humans and machines. In the design of robots today, entire design teams are focused on the user experience and human-robot interaction. So far, although they are starting to talk about it, these teams are not focusing on effective design of robot-robot interaction, especially across platforms and companies. It is still—if it is addressed at all—an afterthought, not only because it is conceptually challenging, but often because it is not considered to be in a company's competitive interests to make their technology play well with others.

That way of thinking is no longer helpful. We need to start by determining when and how we can design people out of the system. We'll

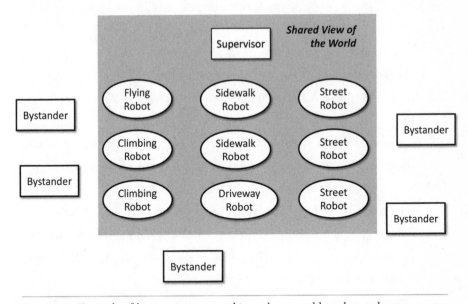

FIGURE 18: Example of how numerous working robots would work together to canvass a neighborhood according to a shared set of goals.

illustrate in this chapter, through historical examples from aviation, transportation, and surgery, that this decision—to design the person out—imposes new requirements on system design. The same considerations will hold for the design of complex robot teams of the future. But now we must go beyond designing for a single robot. We have to determine the appropriate roles for a whole cast of these new social entities that interact with each other, and ultimately, that will require us to design working robots as individual parts in a collective intelligence.

ROBOTS THAT MAKE US MORE CAPABLE BY *NOT* WORKING WITH US

Engineers have been thinking about what tasks may be better accomplished by machines than by humans since the beginning of the digital revolution. Even in the 1950s, human engineering psychologist Paul Fitts developed a list of these tasks.[3] His list was thorough, and it can still be used today to guide decisions about when a task should be fully handed over to a robot:

Timeliness—Decisions and actions that must be made quickly often cannot wait for negotiation between a person and a robot. Robots working together can respond quickly to control signals and resolve conflicts without the delays that people introduce when they have to attend to a situation, collaborate on a solution, and take appropriate action.

Computational Requirements—Machines can quickly and reliably crunch numbers using predefined formulas or algorithms. By contrast, people often introduce errors as they perform mental computations to transform available information or control inputs into a decision or an action, especially when they are under stress. In addition, planning for coordination among many entities—as will be necessary to orchestrate the work of fleets of delivery robots—is something that's very difficult for people to do, but that machines excel at doing. Because robots are good at optimizing decisions across many dimensions and many different variables, computational tasks should be relegated to them.

Reliance on Short-Term Memory—Robots can store and retrieve information more reliably than people can. There are limits to the mental "scratch pad" of humans—that is, the various things being held in short-term memory at the same time. In the robot delivery example, that means the supervisor might not be able to remember which robot is delivering which packages and in what order. But each robot can easily store this information, update it as they make deliveries, and communicate that information across the fleet and back to the command center.

Repetitive Tasks—Because the steps are well defined, repetitive routine tasks can more effectively be executed by robots than by humans. The automatic transmission is a great example. It relieves the driver from shifting gears manually, offloading a tedious task from drivers so they can focus on other things.

Simultaneous Tasks—Human performance declines when we try to multitask. Texting while driving is an extremely dangerous activity with grave consequences. Machines, however, can perform many tasks at once without their performance suffering. Computers are designed for parallel processing, and routines can be scheduled such that the most important task is executed first. As teams of robots coordinate to accomplish tasks, multitasking protocols will become even more important.

Impaired Decision-Making—There has been a surge of interest in robot features that address situations where people are impaired or their attention is diverted at critical moments, causing accidents to ensue. When operators of vehicles and other types of equipment are drunk, distracted, sleepy, or unconscious, tragedies can result. Robots could be used to identify when a person may be impaired, and thus less capable of effectively interacting with them, and adjust their control strategy based on that understanding.

Modern aircraft systems are one of the early examples of needing to purposely remove a person entirely (in this case, the pilot) from interacting with it in certain ways, leaving an autonomous process to take over (here, in order to fly safely). Early aircraft were designed with stable handling characteristics, so that when a pilot pushed on a control and then let go, the aircraft would naturally settle back to steady flight. In this simple way, the airplane was designed to be an ideal collaborator with the human pilot. The pilot was always in control, directing where the airplane flew and how, but the aircraft dynamics ensured that together, the pilot and aircraft could recover from nearly every mistaken human-induced maneuver.

Over time, the goals for aircraft automation became more and more ambitious. The military needed stealth planes that could evade radar. With new analysis techniques, designers changed the classic shape of the aircraft body and wings to reduce its "cross section" observable by radar. There was a problem, though—the aircraft became inherently unstable. The shape and angle of the wings made aircraft less controllable, so that no mortal human could observe the aircraft state from within the cockpit with their own eyes and make just the right precise control movements to keep the aircraft flying safely through maneuvers. To understand the difference between stable and unstable aircraft, think of the behavior of the pendulums you studied in high school physics. A pendulum hanging down will naturally swing back and forth, and eventually it will settle back to center—it is stable. But an inverted pendulum will fall over with just the slightest nudge—it is unstable. To picture an inverted pendulum, think of a grandfather clock—but then imagine turning it upside down. Now, instead of swinging naturally and keeping track of the seconds as they pass, the pendulum has become unstable. It will most likely fall to one side and stay there. Or think of trying to balance a long wooden dowel on the tip of your finger. It takes a lot of practice for a person to keep an inverted pendulum balanced that way—but it's possible. In aircraft, there are complex aerodynamic forces at work on the different surfaces, but in general, the weight distribution of a high-winged aircraft benefits from the pendulum effect, which gives it stability, and in stealth aircraft this advantage has not only been removed, but now half a dozen aircraft control surfaces are acting like inverted pendulums. If you tried the experiment with the wooden dowel, imagine trying to balance half a dozen aircraft control surfaces simultaneously. There is clearly a human limit on what is possible.

Pioneering research in the 1940s through the 1960s, some conducted at MIT, developed the science to model exactly under what circumstances a person could "close the loop" and provide the appropriate control inputs to keep a dynamical system, such as an aircraft or car, stable.[4] Incredibly, research across a large number of studies found that, with some practice, people develop a "feel" for naturally adjusting the actions they take (i.e., their control input for the system), so that the overall system behavior mimics the behavior of a stable system. In other words, as far as they are able, humans naturally fill in the gaps to ensure that the system behaves well. And they do so in a predictable way, which is well modeled by mathematical equations.[5] This model can be used to study whether a human operator, given the constraints that limit their behavior, can control an aircraft, car, or other system.

The constraints on human behavior include, for example, individual reaction times, and whether the person can directly control the position or velocity of the system, or only its acceleration. But there are, of course, a number of other "human factors" that can also limit someone's responsiveness. There is a psychophysical limit on the amount of time a person can scan a gauge and understand its readout, such as for the current attitude of an aircraft. "Noise," or potential errors in a person's observations of the system, or in executing the intended actions, also affects performance.

So the question becomes, How do you design an autopilot system to ensure that pilots—with all their limitations—can maintain control over the airplane, particularly when it's inherently unstable, like a stealth jet? The answer is that more of the control effort has to be shifted to the machine—such that the human's input no longer directly controls the aircraft. Instead, the aircraft takes the human's input and computes the precise behaviors needed to accomplish the spirit of the pilot's directions while flying safely.

This may seem like a risky proposition—something that surely would be limited to military applications. But in fact, many modern commercial aircraft also operate with relaxed stability. Partly, it's for the benefit of reduced fuel consumption. But as passenger jets got bigger, designers also moved the wings lower down, and the engines higher up, to make room for the turbofans.[6] The flight control computers were designed to pick up the additional workload, such as adjusting the aircraft elevators, so that the aircraft could behave as if it had better longitudinal stability than it actually did. This took place with some of the Boeing models, including the brand-new Boeing 737 Max, which was grounded in March 2019 after two crashes within five months.

The problems that led to the 737 Max accidents are still under investigation, but reports to date indicate that a sensor failure led to an incorrect action by the Maneuvering Characteristics Augmentation System (MCAS), a flight control law used in manual flight to enhance the pitch stability of the airplane and make it handle similarly to previous 737s, even though the engines are lower. It activates when sensors indicate the vehicle is entering too steep of a climb and forces the nose down.[7] Reports indicate that there were conflicting sensor readings that caused the automated system to push the plane's nose down when it shouldn't have because there weren't enough failure checks built into the system. So while the system made a challenging aircraft easier to fly, it also led to catastrophic crashes, because the pilots didn't understand what MCAS was doing and were not well equipped to identify and compensate for the failures. Despite the crashes, the concept behind the automated system is valid, but the situation highlights again the challenges and importance of designing for effective human-automation interaction. Aerospace engineers will no doubt learn from this mistake, and we will once again see a major improvement in safety as the learnings set in. The automated systems in these aircraft can make pilots more capable—capable of flying an unstable aircraft—by ensuring that they have to do less.

It's a bit counterintuitive, but the idea is similar to the automatic transmission in your car. Your car can optimize the shifting of gears to minimize fuel consumption much better than you can when driving in manual mode. You still direct the car to drive forward, or in reverse—but much of the workload of striking the perfect balance among acceleration, fuel consumption, and passenger comfort is offloaded from you to the computers in your car. Now, a young person without any experience driving can get behind the wheel and learn without having to struggle with the clutch and gear-shifting—and experienced drivers can devote more mental space to figuring out the newfangled, complex entertainment system. And sometimes your car makes you safer by removing you from part of the controls at just the right time. Remember antilock brakes? When the car senses it is skidding, it chooses to ignore our panicked response of stomping our foot on the brake pedal. Instead it transforms that input into a gentle pumping of the brakes to slow the car while maintaining traction.

Trains—and even surgical robots—use similar strategies, setting boundaries around our control over the system to keep us safe. Modern high-speed trains in France, Germany, and Japan constantly monitor their distance from other trains, limiting how fast the engineer can push the train by ensuring a safe separation.[8] The same technology makes it possible for surgeons to perform seemingly impossible feats—such as operating on a beating heart. No mortal surgeon could cut perfectly along the surface of the heart while it is in motion. Surgeon-robot collaborations, however, make it possible. The surgeon sits at a console to remotely operate a surgical robot. The robot takes care of the humanly impossible feat of "bouncing" the scalpel synchronously with the motion of the heart, and the surgeon directs the cut on the heart's surface.[9]

The first question in the design of such a system is how to safely remove people from the control loop. It's not as simple as designing them out completely. *How* you design them out is equally as important. There are several options, and each will have a different result on overall human behavior, and therefore on human-system performance.

As pioneers studying the intersections between humans and automation have said, "Automation is not all or none, but can vary across a continuum of levels, from the lowest level of fully manual performance to the highest level of full automation."[10] Researchers have proposed "Ten Levels of Automation" to define the full range of possibilities, from the times when a person is in complete control, to the various ways of sharing control, to the times when the machine is in complete control.

TABLE 2

TEN LEVELS OF AUTOMATION

1 The computer offers no assistance, human must take all decisions and actions, or

2 offers a complete set of action alternatives, or

3 narrows the selection down to a few, or

4 suggests one alternative, or

5 executes that suggestion if the human approves, or

6 allows the human a restricted time to veto before automatic execution, or

7 executes automatically, then necessarily informs the human, or

8 informs the human after execution only if the human asks, or

9 informs the human after execution if it, the computer, decides to, or

10 decides everything and acts autonomously, ignoring the human.

Source: R. Parasuraman, T. B. Sheridan, and C. B. Wickens, "A Model for Types and Levels of Human Interaction with Automation," *IEEE Transactions on Systems, Man, and Cybernetics, Part A: Systems and Humans* 30, no. 3 (2000): 286–297, https://doi.org/10.1109/3468.844354.

These options focus on who makes the decision and on how the decision is communicated to the other party. But as cases arise more and more where we need the robot to make the decision, and the user is removed in part, we'll need to take a closer look at how to expand the interchange between the user and the robot for specific tasks. If the robot is going to make a certain type of decision, shouldn't you make sure it will tell the user about the action it is going to take, or, indeed, is already taking? Should you allow any high-level input from the user? This leaves us with three variations for designing the user out of the control loop, as shown in figure 19.

In each of these cases, the robot is responsible for acting on the world in some way. The first option, *input only*, essentially makes the

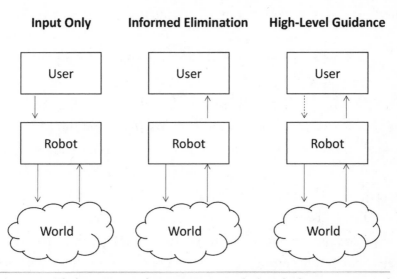

FIGURE 19: Three variations of supervisory control when the human is removed from the control loop: (1) providing high-level input only, but not receiving any feedback on the robot's selection and execution of fine-grained actions; (2) providing feedback without the ability to provide input; or (3) providing high-level guidance to the robot and delegating the selection and execution of fine-grained actions.

human a consultant to the robot's decision. A user gives an input to the robot, but ultimately the robot executes the task without providing feedback.

But why wouldn't you provide feedback to the user? If we don't want users to take an action because they may cause harm, then we may need to be sure not to have cues being presented back to users enabling them to take additional actions based on this information. This approach eliminates the user's ability to intervene even in an informed manner, because the user will not have any knowledge of the robot's actions. This option must be chosen carefully—and only when human input is truly not ever expected to be needed—because it withholds from users the knowledge they would otherwise need to step in and help if the robot malfunctioned. In other words, it only makes sense to go this route if you have good reason to believe that the risk of a human intervening in a system and messing something up is greater than the risk of a robot malfunctioning.

An everyday example we have already discussed is antilock brakes. The driver pushes the brake pedal, but the antilock brake system, detecting when the wheel is rotating significantly slower than the speed of the vehicle, reduces the force on the brake to prevent the wheels from locking up. Antilock brakes detect and respond to this condition much faster than a driver is capable of doing. Recall that the early antilock braking systems alerted the driver about their activation via vibration through the brake pedal. Drivers were confused and alarmed by this unexpected signal, and many responded by taking their foot off the brake, which was not the correct response.[11] The vibrating feature was subsequently removed to mitigate this incorrect response. In this case, the option of taking control away from the user entirely was appropriate, because the alternative was causing drivers to take an unsafe action. Managing the pace of the deceleration was best left to the antilock brake system, and the driver truly did not need to know about the system's choice.

The second option, *informed elimination*, fully removes the user from the action sequence, but makes this clear to the user, and informs the user of the actions being taken. An everyday example would be a driver-assist feature called a *collision avoidance system* (CAS).[12] CAS uses a set of sensors positioned around a car to monitor for signs of a potential crash. If it detects such a hazard, CAS first warns the driver, giving the driver a chance to take action. If the problem worsens, then CAS takes corrective action, by braking, by steering, or both. In this case, the driver is aware of what the automated system detected and has a chance to solve the problem, but the system can also act independently and override the user.

The third option, *high-level guidance*, accepts user input, but treats it as guidance toward the user's intent. The robot transforms this guidance into what it considers is the right action. This type of system is appropriate when you expect users to provide general direction, but users are not capable of executing the detailed sequence of actions themselves as quickly or reliably as a computer can. This design approach is different from "input only" because supervisors continue to monitor the detailed execution of the action, and are therefore able to continually update their input. It has been used in the design of spacecraft, stealth jets, and many other applications. Here, users provide the high-level intent in a way that is intuitive to them, but the system determines how to achieve that intent.

When the user/supervisor cannot execute a given set of detailed steps in a timely manner, this type of design is useful. We see that principle at work in the remote operation of robots from a command center, for instance, because there can be latencies in the communication path that delay a supervisor's input to a robot for seconds or even minutes, depending on the network characteristics, making it impossible for a supervisor to take direct control of the robot and drive it.[13] After providing guidance and monitoring the robot's execution of the decision, the supervisor can decide to intervene again, if needed.[14] For example, if a robot approaches a construction zone and is unable to find a path around it, the robot can ping a command center for help. A supervisor can then plan a safe path around the many cones and barricades that aren't easy for a robot to understand, and the robot can safely follow the path provided. This is how the Mars rovers have been controlled.[15] It takes a long time to communicate between Mars and Earth, so it's not feasible to have someone sitting in NASA's command center controlling *Curiosity* with a joystick. Instead, the rover's handlers on Earth send it detailed guidance on what path to traverse, and it then executes this plan on its own.

The systems we've described thus far involve relatively simple interactions between one human and one machine (the pilot and the aircraft, the driver and the car, the engineer and the train, the surgeon and the surgical robot). In these examples, the human's effect on the system, as well as the environment the system is operating in, are both predictable. The wisdom we can take from the aviation industry is particularly limited, because our airspace is a much more predictable and less complex environment than the one we encounter in our everyday world. Wherever we may go, the situation on the ground is more complicated than the situation in our airspace: there are more people, they are all doing different things, and they are doing them without a common purpose.

As fleets of delivery robots invade our neighborhoods, they will join a world where dogs are barking, children are playing, cars are driving past, and road conditions change with the weather, sometimes at a moment's notice as rain clouds burst or snow or hail starts to fall. The computers on board the robots will be ready to communicate with each other, but they won't yet know precisely what they'll encounter on any given day. Their interactions with the environment and with other people will be almost wholly unpredictable. It will be too much for one person in a delivery truck command center to fully monitor or control.[16] The question is, How and when do we limit or expand the information presented to supervisors, and how and when do we circumscribe the control these supervisors can wield over the system, so they can oversee the safe and efficient delivery of your packages? And how and when should the system let bystanders intervene, whether it's to direct robots away from their children or to get them to go around their landscaping projects?

MANY PEOPLE, MANY ROBOTS

If users can be designed out of the control loop for certain tasks, then how do we support robots in completing those tasks independently? It's not only about one robot's tasks: we also need to support the coordination of tasks across many robots. As robots enter every aspect of our everyday lives, the design problem is much more complicated than the interaction between a single robot and a single person. We must design robots for a world where they will need to collaborate with *many* other robots while also interacting with people. How will they decide when and how to partner with supervisors and engage with bystanders?

Taking the example again of a fleet of robots collaborating with each other to make deliveries in a neighborhood, now consider that there would likely be other robots operating in the same neighborhood simultaneously. But these other robots wouldn't be working for that same delivery company. They might be delivering dry cleaning, taking pictures for a new house listing, escorting a child home from school, or carrying groceries, not to mention the many autonomous cars or trucks navigating the streets. These robots will intersect with each other at crosswalks, on sidewalks, on people's driveways, and even in doorways. If they aren't able to collaborate directly with each other, then many interactions will end badly. There will be collisions, and there will be robots getting stuck, like the autonomous cars that detected each other at the intersection and couldn't negotiate with each other for forward progress.

The robots developed by any one manufacturer (e.g., the fleet of delivery robots) will presumably be operating with a shared view of the world, with the ability to directly communicate with one another. Because they were developed by the same design team, they will be able to strive for optimal understanding across their team of robots and supervisors. But the robots that were developed by different design teams and manufacturers may be different enough that the chasm for communication will be too wide. How can these robots, at a minimum, coordinate with each other during moments of intersection, or even coordinate with each other at a broader scale, so they can benefit from each other's knowledge and experiences? If they could share their experiences and combine their knowledge, then suddenly we would have a network of machines that were all imagining the world in the same way at once, and each machine (and user) would get to reap the benefits of what every robot knows.

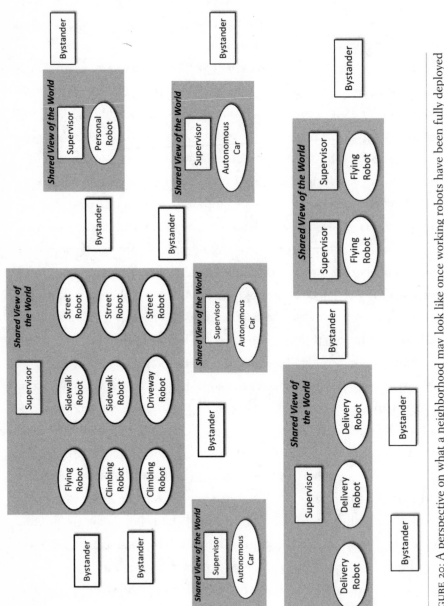

FIGURE 20: A perspective on what a neighborhood may look like once working robots have been fully deployed across many commercial applications.

Researchers are actively working to address the challenges that come with fleets of robots working together to achieve common goals.[17] But these approaches still assume that the robots are working as a team toward a common set of objectives, and that their activities are highly coordinated. The tasks, including which agents can do what, and how well, are modeled explicitly. They characterize uncertainty mathematically, where some particular task performed by a particular agent has some probability of success. Because the robots are working toward a common objective, that objective is precisely codified with some notion of "reward" accrued to the robots for successful performance. The existing research on this problem assumes that all agents share a common understanding of each other's priorities and subgoals, as well as how their subgoals may interrelate. These algorithmic approaches also typically require that the agents share many communications about what they observe, what they have just accomplished, and their progress toward the goal. This type of approach is useful for tasking many robots with, say, search-and-rescue operations, because it allows them to coordinate the search as they visit different parts of the search area at different times. And it works when robots are designed to talk to each other directly, when they have appropriate bandwidth to support these communications, and when they have the right computational capabilities to process the communications and make sense out of them.

The existing research in multi-agent collaboration does not, however, often address the interaction between robots on different teams that still need to coordinate with each other. We need some new ideas. We can draw on a number of different approaches that have been developed for multi-agents working on the same team. But we have new problems to solve to achieve safe and efficient cross-team interoperation. We need answers soon, because fleets of different robots are coming to our dynamic human world.

We have scarcely conceived of a system for doing this, let alone implemented one. We can sketch out here what such a system would have to do in order to work properly. There are specific needs when it comes to robot interoperation:

- **Micro-interactions:** Since robots from different companies are not operating as a team, they won't know of each other until their activities intersect in some way. We will need a way to facilitate micro-interactions when these intersections occur, and we will need to be able to deconflict their activities.
- **Negotiation:** Since robots will have their own decision systems and reward structures, we will need a way for them to negotiate and resolve conflicting objectives without knowledge of each other's objectives.
- **Robot crowdsourcing:** Since robots will never build or maintain a complete view of the world in which they operate, they will need to crowdsource and share information with other robots and people, similar to how Waze crowdsources traffic information across drivers in a manner that preserves privacy and security.

Let's examine each of these in turn.

OUR STREETS ARE BUSY, BUT THEY WILL BE EVEN BUSIER WITH ROBOTS. Currently, there are few rules that might be adapted for guiding their behavior when they get to the point where they will need to interact with other robots or with people. So we are starting from scratch. How will these new social entities maneuver around one another and around us without doing harm? One active area of research studies how birds flock and how fish school, looking for inspiration for how robots might move in elegant patterns. The birds and fish do not seem to be directly communicating their movements to one another, and so maybe our robots don't have to, either. A flock or school forms when each individual bird or fish simply tries to maintain the velocity vectors of the bird or fish next to it. This very localized type of coordination, however, is essential to the collision-free motion of the entire group. As long as each bird or fish flies or swims parallel with the one next to it, they will all fly or swim in a harmonious swarm without running into each other. In robot motion, this would mean that it might be enough for a robot to be able to maintain a tracking distance and follow the direction of its nearest neighbors. The robots could avoid collisions in this way—and even engage in basic cooperative behaviors. Consider ants, with their complex-seeming foraging behaviors. Ants chemically mark their successful trails as they forage. Even simple ways of sensing the world and communicating can help these tiny representatives of the animal kingdom realize a myriad of complex behaviors.[18]

Why does this simple strategy of leaving a chemical signal work so well for coordination across an ant colony? An ant doesn't have much intelligence, so complex communication isn't an option. Nor is there any single, centralized manager directing all of the colony's ants. Robots will be similar, but for different reasons. Communication between them will need to be simple because they will have different processing architectures, and any external communication will need to be fluid. They have different goals and will be designed by different companies, so there can't be a centralized manager.

Could we enable robots in our everyday lives to have effective micro-interactions, such as digitally marking their trails, while they go about trying to achieve their objectives? They would need a way to communicate this digital trail to other robots in the same neighborhood or headed in the same direction. Just like us, robots will need to decide which street to turn down to get to their destination, and often they will have myriad choices. By knowing where points of potential interference with other robots may occur, they could make small adjustments in their own paths to avoid these difficult intersections. The first need for robot-robot communication is simply for them to be able to communicate enough information to avoid colliding with each other. For example, they could communicate location, direction, and speed. In the air transportation example introduced earlier, each aircraft interrogates each other aircraft's transponders. The transponder replies in a way that mimics radar, and based on that information, the Traffic Collision and Avoidance System calculates the position of the other aircraft. If the two aircraft are on a collision course, then the TCAS algorithm calculates and issues a resolution advisory to prevent the collision.

This is purely a reactive conflict resolution function that kicks in only when two aircraft are at risk of colliding. The interesting thing about it for our purposes is that the resolution advisory is calculated based on only two pieces of information—position and velocity—from each aircraft. It does not complicate things by considering intent, objectives, or flight path. This type of approach doesn't require that kind of detailed information, and it could easily be adapted to support the millions of micro-interactions between working robots that will occur every day when they canvass our cities. This is similar to the minimal information sharing that's used by flocks of birds and schools of fish, which is more evidence that micro-interactions can be effectively achieved at scale.

Of course, the complexity of tasks performed by ants, fish, and birds is dwarfed by the complexity of our human world. Ants, fish, and birds don't need to retrieve or deliver their food on Amazon's tight schedules, and the number and types of interactions they can have with their surrounding world are much smaller than the set of possible interactions that robots will need to deal with operating on our streets and in our neighborhoods, interacting with any number of bystanders. Besides, colonies, schools, and flocks are, at base, made up of individuals coordinating toward the same goal. Fleets of robots, operated by different companies, will need to negotiate competing objectives. Nor do we want to see whole fleets of robots heading toward us on the street like an army. As our streets become increasingly clogged with these systems, it is entirely plausible that cities will follow the lead of San Francisco, and put limits in place about the number of sidewalk robots that can concurrently operate within a neighborhood. Robots will not "live" in a single neighborhood in any case: they will be regularly traversing the city. And so they will need a way to plan their collective routes in a way that will prevent collisions, not become too much of a nuisance, and follow a city council's rules, all while meeting their companies' objectives.

Research in multi-robot coordination has developed a number of "bidding" and "market/auction" algorithms for decentralized coordination of robot tasks and routes.[19] As in an auction, in this model a robot must bid an amount according to how important an option is to it. This creates a mechanism to mediate the trading of options such as routes, information, and precedence among robots, without them needing to know one another's underlying objectives. Robots from different companies are naturally disincentivized to share details of customer needs, planned routes, and relative priorities over tasks. These will remain the trade secrets of each company. However, robots can potentially share their relative priorities over access to resources—for example, which neighborhood they prefer to access in what order as they traverse the city, which robot gets first access to use the sidewalk with fewer cracks, and which must cross over to the other side or wait their turn. A number of algorithms exist to take these relative preferences of many robots to converge on a "fair" allocation of the available resources, while still preserving privacy of the full plans and objectives of each individual agent. In the market-based approach, each robot acts as both a buyer and a seller of tasks or options.

The auction is coordinated by an "auctioneer" agent that announces the set of tasks and options available during a particular negotiation phase. Each robot that wants to perform a task or reserve an option (e.g., passage through a particular city block) then submits bids based on its preferences. Each robot must know how it values the tasks or options itself, but does not have information about the other agents' preferences.

The preferences of the robots are reconciled through an interactive bidding and assignment process. This approach eliminates the need for some central authority to maintain global information about robots' private information or objectives, and it is flexible enough to adapt to a situation as the number of robots involved in the bidding process changes over time.

Consider the situation in San Francisco in 2017, when the city determined that no more than nine sidewalk delivery robots could operate within a given sector. One can easily conceive of a dozen different companies offering the services of working robots to the most populated areas of the city—takeout delivery, package delivery, shopping and errand robots. The various companies will have to negotiate their routes, ensuring that the sidewalks are never obstructed for pedestrians. Imagine that each time a new robot enters the sector, it aims to file a "flight plan," reserving a spot in each city block along its preferred route. It can't be sure it will get its top preference for route, so it submits ranked preferences for a few different possible routes. The robot's "flight plan" is confirmed via a decentralized action method and bidding process, rather than through a central authority, like air traffic control. As robots enter and leave the space, and as more robots request entry (during a future robo-rush hour), the robots renegotiate their plans in a fair manner using the algorithmic approach.

The challenge in directly applying the auction algorithms to coordinating multiple fleets of robots is that each robot on its own may have a difficult time assessing its own priorities and preferences over tasks and options. Our world is just too dynamic, and the information that any one robot or set of robots holds about the world soon becomes out-of-date. A robot might believe it is bidding for the fastest route to its busiest neighborhood, but then on the way discovers that a branch has fallen during a recent windstorm, requiring a long detour. Meanwhile, just a few moments ago, a robot from another vendor discovered the same obstacle and began taking the detour. Nevertheless, while companies naturally don't want their robots to share their customer lists, their schedules, or any other sensitive data with other companies' robots, there is no doubt they would all collectively benefit from having certain cooperative agreements to aggregate some of their data about the world. Essentially, they need a robot crowdsourcing platform similar to Waze.

In the Waze app, we share information about accidents, detours, and other things that may benefit other drivers, in return for those drivers sharing information with us. To benefit others, we don't have to share any sensitive information about where we are going and why. Just as we sit in our cars wondering whether the bridge will be backed up today, our robots will sit in their intersections making blind guesses about how quickly they might get to where they are going based on yesterday's data. Unless we give the robots a communication channel, so they can share their information about the world, they will frequently get it wrong. Carefully constructed task-and-option auctions will not result in robots achieving their goals, and the robots will make life harder for everyone around them in the process.

And what about us? Imagine that these robot-robot issues are solved, and that in our future world, robots cooperate through many different means to share information, refine their objectives, and safely and efficiently negotiate our streets, workplaces, and homes. Some of their methods won't even make sense to us, because their objectives are not clear or are prescribed by code embedded in opaque machine-learning algorithms. So what exactly do we need to know about what they are doing and why?

We've discussed the reasons why a person might need to be removed from the loop, and introduced three models for different levels of minimal human involvement: input only from the person to the robot, without any feedback provided; informed elimination, in which the person still provides information to monitor the system's behavior; and relegation of the human to the position of a guide, in which high-level guidance from a supervisor is translated into fine-grained control inputs by the robot. The appropriate model for any given situation will depend on the particular context, such as whether we are considering human-robot micro-interactions while passing on the sidewalk, or negotiation of access to intersections and travel routes.

There is also an exciting new open design space in how we can crowdsource knowledge from people in the world to make the robot collective smarter and more capable. For example, Waze gamifies the process of sharing information. Most people don't volunteer new information because of a conscious awareness that by doing so, the benefits flow back to them. We do it because it's fun! We can think about how to design similar apps that make it fun for people to help the robots. Of course, the benefit will also flow back to us, because our packages will be delivered on time and without the robots becoming a nuisance.

Similarly, we can design apps that enable a person walking down the street to take a peek inside a robot's brain—what they see, what path they are planning, why they just changed direction. Say you just picked your baby up from day care, and you are walking home from work with the baby in the stroller. You see a robot stopped in your path, seemingly doing nothing. And it's quite hard to maneuver the stroller around the robot. You pull out your phone, open your robot crowdsource app, and select for the robot closest to you. From the robot's view within the app, you see that it is a pizza delivery robot, and it's waiting for a person to come out of a nearby apartment building to retrieve their pizza. With a few clicks in the app, you politely tell the robot it is in the way of pedestrians, and point it to a safe location in the walkway of a nearby building. The robot thanks you, and then shares this new bit of information with other robots that will later come through the area. With this small act, you don't have to worry about another robot from a different company blocking your path in that way again. Getting the interface right is not trivial. By leveraging our design approach from chapter 5, we can develop interfaces that make the robot's capabilities transparent to bystanders and directable by them, and we can enable them to quickly calibrate trust in the system.

CHAPTER EIGHT

This City Is a Cyborg

IT'S THE END OF THE DAY ON A FRIDAY, AND YOU DIDN'T MAN-age to make it to the mall to pick up the favors for your child's birthday party this weekend like you thought you could. So you log on to your Amazon account to see what might be available for next-day delivery through Prime. In addition to the favors, you find the lightbulbs you need to replace the burned-out bulb on the table lamp, and also spot a new book in the recommended section that you decide to go for. You click "Place my order," and a short while later, robots in Amazon's ful-fillment warehouses are whizzing away to make sure it's delivered to you right on time.

The warehouse is full of small, flat robots that shimmy underneath shelves loaded full of everything from blenders to wool coats to table saws. Once underneath the requisite set of shelves, a robot lifts it up and transports it around the warehouse. When your order is queued up, one of the robots carrying the shelf with your product is notified. The robot zips through the warehouse, stopping and starting, moving left then right, "dancing" around all the other robots who are also moving through the warehouse. It's a truly beautiful sight.

"These robots are so intelligent and efficient, so much more so than human workers," you might think. And in some ways, they are. But if you were to tour Amazon's warehouses, you would also learn that each robot is mostly blind. The robots have very few sensors. They "navigate" their world by looking straight down at paper taped onto the floor by human workers, known as *fiducials*. The warehouse is one big grid, with a unique paper pattern taped to the floor of each grid square. The robots simply track the sequence of paper patterns they pass to confirm their location. Sometimes one of the pieces of paper is ripped up by the robots' wheels, and a person pauses the robots to walk into the space and retape the paper to the ground. Amazon has 175 fulfillment centers around the world, and these robots currently zip around in 26 of them, with humans and robots working together to fill your orders.

This may sound like a hacky solution, developed by a start-up to get the robots out the door. But Amazon, one of the most successful companies on Earth, still chooses to tape the paper to the floor. Why? Robots with fewer sensors are less expensive than robots with more, but more importantly, with this approach they are less likely to fail. Robots are much more reliable if you program them to follow a pattern of papers on the ground than they are if you try to make them capable of observing the world around them, detecting obstacles, planning a path around the obstacles, and then continuing to look for their destination. In other words, sometimes simple is best. The working robots of the future will be expensive, complex machinery, but ultimately, they will still need our help, and sometimes, that will require more of us than just giving them directions. Our societies are not currently built to handle the needs of independent robots, and it's not clear that we can simply make robots that only need what our infrastructure currently offers. We will need to make some changes. The good news is that we can change our environment in small, simple ways that will make the world much easier for our limited robots. In this chapter, we will explore analogies for taping paper to the floor that will enable working robots to reliably navigate our society. How can we think about changing our environment and our infrastructure to pave a way toward a future that seamlessly integrates these new entities? What big and small changes can we make to the "rules of the road" that govern society? And given that we cannot transform our neighborhoods, streets, and stores in a matter of days or weeks, the way a warehouse or a factory can, what is our way forward to realizing the bigger changes that are needed?

Our environment is built to organize our actions within it. We have traffic lights, interstate on-ramps, and pedestrian crosswalks to help coordinate driver and pedestrian activities safely and efficiently. We mark doors "Exit Only" to keep people from entering buildings in the wrong places, and numbers on tracks in train stations to help us make our way to the right train. Speed-limit signs are methodically placed along the highway so we know the safe speed as we transition from urban streets to highways. Orange cones line construction sites to communicate that dangerous activities are in progress and roads or sidewalks may be closed. Crosswalk guards stand beside busy streets and hold up their hands to signal for cars to stop as children cross the street. Robots need the same type of structure and support, maybe even more. A robot's sensory systems are disadvantaged compared to ours in many ways. They take in lots of data about the world around them, but they are not as good as we are at deriving meaning from all that information. This also means the kind of support they'll need from their environment isn't the same as what humans need. But what will it be, then? To inform our thinking, we can look to how industrial environments have been redesigned to support advanced machines.

THE AVERAGE AIRPLANE PASSENGER PROBABLY DOESN'T REALIZE THAT THE plane he or she is riding in is flying in a lane. This virtual trail of bread crumbs, designed around a network of fixed ground beacons, revolutionized aviation safety. These navigational aids help the pilot and air traffic controllers track the plane's location and have been in use since long before GPS existed. Additionally, our airspace is divided into different flight levels and tracks that act like lanes on a highway, except these lanes are very wide (and tall) to accommodate the high speed of aircraft, potential errors in location estimates, and other factors, such as the wake vortices created by each aircraft. Today the vertical lanes are separated from each other by one thousand feet. These lanes in the sky minimize the potential for aircraft to cross each other's paths unexpectedly and collide. They also simplify the procedures for managing air traffic. For example, if two aircraft are on a collision course, rather than trying to calculate exactly when each one will arrive at the collision point, or recommending a slight maneuver to one of the pilots to prevent the collision, air traffic controllers typically ask one to climb one thousand feet—and then the two planes are guaranteed not to come close to each other, or to any other plane in the sky, because they are in separate lanes.

Structuring the airspace in this way has had a tremendous impact on the efficiency and safety of air transportation, because it offers clear rules that regulate the behavior of every plane in the sky. It has also made air travel much more efficient, because it's organized. As technology has evolved and more navigation capability has been built into each aircraft using satellite navigation and automated position reporting, a new, more adaptive approach has emerged called *performance-based navigation*.[1] As long as an aircraft has navigation capability that is equal to or better in performance than the beacon-based approach, an airline can apply for more flexible routes that deviate from these lanes, thus improving both overall flight time and fuel efficiency. Air traffic control procedures have adapted to maintain safety for these higher-performing aircraft. For example, areas of high traffic, such as airspace near airports, are so busy that they require strict, consistent organization.

So even more capably equipped aircraft are still managed with well-defined tracks.

In factories across the United States, Europe, and elsewhere, we structure and organize our spaces for robots as we do for commercial airplanes. And the ways we do this for robots are evolving and becoming more flexible just as in aviation, as robots gain sensing capabilities and intelligence. Fences have usually been built around robots to physically ensure that they will not pose a safety hazard to human workers. But this model is disappearing quickly as technology evolves and new safety measures are developed. Scientists have created new, dynamic ways of marking "personal space" for people and robots, and this enables close physical collaboration to take place in manufacturing without putting workers in danger.[2] In place of a static demarcation of robot and human space, the industrial environment is outfitted with new sensors that function effectively as virtual fences.

In the new system, if a person moves close to a robot and crosses the virtual fence, the robot immediately stops moving, to ensure it does not accidentally make contact with that individual. In the more advanced versions, the sensors are used to create dynamic safety zones, in which the distance between the person and the robot is actively monitored. As the person nears the robot, the robot slows, giving the person time to react before the robot stops completely. Think of it as two people passing in a tight corridor: you each slow down as you near each other to ensure that you can maneuver out of each other's way in time. This simple switch, from physical to virtual fence, and from static demarcation of safe space to dynamic adjustment of safe zones, provides for new methods of human-robot collaboration in manufacturing tasks, and this collaboration allows the two to complete these tasks more efficiently or with higher quality than either human or robot could achieve working alone.[3] The concept of organizing, outfitting, and adapting the environment and tasks to ensure that robots do not pose safety hazards to people has made this reality possible in factories. Air travel and factory work are the precursors for many more such partnerships to come. In the future world, we will coexist with countless intelligent robots, but they will be more advanced than the ones we have today. They will evolve.

We have already learned lessons to chart the path to a collaborative society, and the history of aviation, again, is instructive. From aviation's earliest days, airplanes needed to be carefully coordinated.[4] At first, we simply used bonfires at the end of runways to signal the runway location at night, and pilots looked for the blazing light to find where they needed to land. Next, beacons were installed, and aircraft could use radio navigation to find the runway even on a cloudy day. Transmitters broadcast a modulated signal, which is received on the aircraft. The position of the transmitter is calculated using time of flight between received signals, and this data is used to determine the position of the aircraft. Initially this was calculated by hand; now it is automated, and improvements over decades have made this approach extremely robust. World War II brought us radar surveillance, allowing air traffic controllers to track planes during flight without relying on transmitters, especially in congested airspace such as around airports.

But the real revolution in air traffic came in 1956, following the collision of two planes over the Grand Canyon.[5] The planes were operating in uncontrolled airspace, where pilots are expected to "see and avoid" other aircraft without any external help. Both pilots were maneuvering around scattered cumulus clouds to get a better view of the Grand Canyon, and they both entered the same cloud, making it impossible for them to see each other. They collided while still in the cloud, and all 128 passengers were killed.

This midair collision, which took place during the rise of commercial aviation, created a panic. Aviation rules at the time had no good way to protect planes against these types of conflicts. The solution was to centralize the management of airspace. In the United States, Congress appropriated $250 million for an upgrade of the nation's airway system. To oversee that new system, it created the Federal Aviation Administration (FAA), which was given broader authority to combat aviation hazards than the predecessor organization, the Civil Aviation Agency—which shared authority of the airspace with the military—had been given. The FAA implemented extended positive separation between aircraft and planned "Super Skyways" to connect major East and West Coast cities, carving out airspace with separate rules to facilitate heavy cross-country travel.

This later evolved into the current system that organizes our skies. It includes rules for separation (lanes in the sky), requirements for pilot training, and safety equipment on board. Areas with a busier airspace, such as around large urban airports, have stricter rules. And airspace in the highest demand, such as the North Atlantic routes that get maximum tailwinds from the jet stream, are structured as tracks (similar to interstates) to provide for greater access while maintaining safety.

The first FAA administrator, Elwood Richard Quesada, was appointed by President Dwight D. Eisenhower with the authority to "regulate, establish, operate and improve air navigation facilities; to prescribe air traffic rules for all aircraft; and to conduct related research and development activities."[6] Fundamentally, what was needed was some mechanism of cooperation over the shared resources of aviation, including airspace, and that was best administered through a central organization. Do we one day need such a central agency to create rules, develop external navigation support, and regulate other aspects of robot operation and control? Possibly. But at the very least, industry cooperation for negotiating the very personal shared resources these robots will utilize—our roads, sidewalks, hallways, and aisles—will be key.

Having lanes in a sky seems like a good idea for airplanes, but is this idea of much use for a robot that needs to pass you in a crowded supermarket aisle? There presently is no room for separate "lanes" for robots and people: people can barely pass each other pushing their large shopping carts as it is. Although robots have only recently been unleashed in the grocery aisles, our factories have been puzzling over this problem for a few years now.

Traditionally, robots in factories have operated in spaces separate from people. An automotive factory today is full of big, fast-moving robot arms, but they operate in cages. The robots are kept separate from the people because they can only operate in highly controlled environments. The parts need to be placed precisely—if they are out of place by even a few millimeters, the entire operation churns to a halt. And the robots can't sense people nearby. If someone were to enter their space, it would be a significant safety hazard.

However, the truth is that relatively little work in most factories—even automotive factories—can be structured so carefully for robots. A car body can be built almost entirely by robots, but the rest of the job—installing wiring, seats, and dashboard elements—is still done almost entirely by people. Even modern-day automotive plants have rows and rows of people building up cars by hand, as far as the eye can see. This work can't—and won't—be performed exclusively by robots for the foreseeable future, because it requires skills that robots do not yet have. But manufacturing engineers are realizing that the work robots do already, such as assembly, welding, and packaging, can be done better and faster if robots are freed from their cages to work alongside people. Rather than trying to reproduce the tasks of a human worker, robots freed from their cages can actively assist the human worker—handing over the right part at just the right time, for example—and thereby drastically improve the productivity of the line.

Companies are therefore revisiting the challenge of managing these complex machines in a way that is safe for the people who surround them. The robot smarts required to monitor the progress of humans and anticipate what they need is a far cry from the blind standard industrial robot in a cage, or even from the robots in Amazon's warehouses that navigate using paper markers taped to the floor. Much like the modern complex choreography of planes crisscrossing our skies, factories today need technology that allows for a more intimate, close-proximity dance of humans and robots. And to make this technology work, we need fail-safe methods of ensuring that robots can't harm their coworkers.

The current solution, as specified by the International Organization for Standardization, is to implement something called "speed and separation monitoring."[7] Just as aircraft have different rules for separation in the air, industrial robots must maintain specified distances from people based on their speed. The faster the robot is moving, the farther away it must stay from a person, and as a person nears the robot, it must slow and stop. One of the first systems of this kind was deployed in a BMW plant in Munich in 2017. A human associate worked underneath a towering orange industry robot, two to three times his height, as they safely negotiated shared factory floor space to build cars.[8]

In 2016, collaborative robots represented only about 5 percent of the robotics market.[9] But our studies, and others, show that with close-proximity collaboration between humans and robots, tasks can be accomplished much more efficiently—up to 85 percent faster, according to some of our studies—than when humans perform assembly tasks without robot assistance.[10] As a result, in 2014 the collaborative robot sector was forecast to grow tenfold by 2020, with a market reaching more than $1 billion annually, compared to $95 million in 2014.[11] The takeaway is that the "rules of the road" for working with robots don't have to be static. They can adapt over time as robots become more capable and we become accustomed to them. As robots evolve, we can take them off of their fixed "lanes." Through more dynamic negotiation of shared resources, we can take some big leaps toward integrating robots into human environments.

In many ways, a city sidewalk is much more challenging for a robot to navigate than our airways, and the set of possible human interactions on a sidewalk is an order of magnitude more complex than on an assembly line. There are potholes in the road and cracks on the sidewalk, whereas the industrial setting is well maintained. There are bicyclists whipping past and pedestrians walking close to each other without even noticing as they read the news on their phones. Our lives are full of uncoordinated, dynamic entities whose movements are impossible to predict. The only way to get the technology right—and this is true for many of the changes we've advocated in this book—is to start simple and improve over time. The same is true for design of the environment. As robots start out with less capability and are less proven, we may need to ensure clear separation between robots and people, the way we separate cyclists in bike lanes and pedestrians on sidewalks from the roadway. As the technology matures, we may be able to move toward more flexible approaches, but with clear conventions. For example, semitrucks and cars share the same roadways despite potential visibility and safety issues, but the semitrucks, by convention, will stay in the right lane unless passing. Less interleaving of cars and semitrucks reduces the opportunities for accidents. Already, high-occupancy vehicle (HOV) lanes often change direction with the flow of traffic, from the morning to the evening commute, to help decongest crowded roadways. In "smart cities," such as Pittsburgh, the traffic lights adjust their schedule based on the flow of traffic. Similarly, our robots could rely on smart environment design to cue appropriate behavior in different contexts until they are able to operate without this assistance.

A FUTURE WORLD DESIGNED FOR ROBOTS

Previously we've discussed how to make robots more predictable to people, but we also need to make society more predictable to robots. In other words, we need to give them the same kinds of cues we give to other people.

But first we need a way to frame the amount of chaos or disorder that robots will experience as they are unleashed. Let's call this *societal entropy*. The more the variables change in the environment in which a robot is operating, the higher the entropy, and the greater the challenge for the robot to operate successfully. For example, lanes on the streets and sidewalks rarely change, but the behavior of a dog walking down the sidewalk, sniffing bushes and dropped trash while attached to a leash, is very unpredictable and thus presents a very challenging obstacle for a robot. When things in the environment change frequently, they present more surprises to robots, putting responsibility on the robot designer to cover an unimaginable set of situations the robot may someday encounter. Societal entropy can be measured as the rate of change for each entity (i.e., how fast entities are changing their state around the robot). For the purposes of this discussion, entities can include infrastructure, objects such as people and dogs, buildings, and even rules of the road, such as speed limits. The techniques used in our skies and factories aim to reduce disorder by providing structure for robots, and ultimately by simplifying their tasks and the technical capabilities needed.

We can break down societal entropy into three levels:

1. **Low Societal Entropy:** This level includes our most stable aspects of society, the things that provide the basis for organization and order in our public lives, such as infrastructure, laws, roadways, buildings, rooms within buildings, and speed limits. These are all difficult to change and require investment, planning, and energy to modify. Therefore, they change infrequently, and robots can benefit from that stability.

2. **Midlevel Societal Entropy:** Some things change periodically, and sometimes frequently, but the change itself is predictable. This includes traffic patterns related to rush hour, road closures due to construction, and speed limit changes in school zones. As long as you know the rules that govern their behavior, you can easily adapt to midlevel entropy.

3. **High Societal Entropy:** We encounter highly dynamic and unpredictable entities constantly. These include passing cars or bicycles in front of you, a fallen branch across a sidewalk, and a

road closure due to an accident. These changing forces can create frustration even in our own everyday lives and are by far the hardest challenge for working robots. We rely on a vast library of norms to successfully adapt to them, but robots have not yet fully conquered this level.

Societal entropy is impossible to eliminate. Nevertheless, there are ways we can systematically try to reduce it in order to make environments more predictable for robots:

- **Organize** society with new rules that would benefit robots just as we benefit from our governing principles today, such as lanes on the roads and even common etiquette—like passing on the left on our sidewalks and staircases.
- **Outfit** society with signs and signals that are as reliable to robots as street signs and traffic lights are to us today.
- **Adapt** society dynamically based on the context, just as we do today in challenging situations such as major sporting events or construction sites, where we might, for example, add new rules for cars and pedestrians.

Organize Our World to Help Robots

We can help robots coexist with us by organizing the space in which they operate. Human society relies on a predictable organization of space and on clear rules and procedures governing how to behave. On the highway, the lanes are clearly marked, we know to pass on the left, and there is a lane dedicated to carpool traffic. This organization of space simplifies what would otherwise be an overwhelming task—avoiding countless cars all traveling in close proximity at high speeds. Instead of watching out for cars zooming around us in all directions, we simply stay in our lane and pass on the left if a car in front of us is going too slow. If you've ever found yourself driving on an old stretch of highway where the lane markers have faded, you've probably come to appreciate the way lane markers organize our roads.

In order for robots to successfully navigate uncontrolled public spaces, we will need to incorporate their needs into the way we organize society. We can ensure reliable, safe robot operation if we can find ways to make space for them—literally.

In certain situations we will have the luxury of creating dedicated lanes for robots on streets or sidewalks, and we will need to do this for some robot integration. Our interstates will need a lane dedicated to autonomous traffic, similar to HOV lanes, especially if we want autonomous semitrucks to travel at sixty-five or seventy miles per hour without becoming serious threats to our safety, and particularly when they are first introduced. We may even need a physical barrier between the autonomous vehicle lane and the other traffic until this technology is proven to be safe. It could require substantial investments in our infrastructure, but at the beginning we can test solutions by sharing and repurposing existing infrastructure in a small sample of times or locations. For example, existing HOV lanes could be used by autonomous semis for certain time periods each day.

We will need to accommodate robots in new ways, too, such as with "robot crosswalks" in areas of heavy robot traffic—for example, where they enter and exit a stockroom at the grocery store. We will need to train our toddlers to stop at these robot crosswalks and take them as seriously as we take crosswalks elsewhere. This type of demarcation would let people know where they need to have heightened awareness of the possibility of robots around them. It will also create areas where robots get the right-of-way, which will be necessary for cooperative coexistence. Robot lanes and crosswalks would provide for areas of low entropy within an overall environment that has high entropy.

Of course, any change to our environment comes with hard choices. Who will be responsible and bear the cost of installing these lanes? Who stands to benefit? Surely, we cannot retrofit every sidewalk and store as a precondition for these robots to be successful. And where we install these robot crosswalks and lanes—in what neighborhoods and in which stores—may drive differences in who has access to new working robots, who can benefit from them, and who is impacted by the growing pains of introducing a new technology at scale. We will need to be intentional in determining who will make these decisions, and through what process. We may decide there are benefits in the near term for robots to use existing sidewalks or bike lanes that are well demarcated—but this decision, too, would have implications for robot design. A robot that conforms to the social norms of maneuvering down a sidewalk is very different from a robot that behaves more like a bicyclist in his own lane.

Another option might be for our neighborhoods to work with companies to agree on robot delivery windows, where a package truck unloads a fleet of sidewalk delivery robots to crawl our neighborhoods only from, say, 10:00 to 10:30 a.m., and again from 7:30 to 8:00 p.m. We would know to keep our children and pets inside—and avoid going for a run—during these hours, reducing the societal entropy for robots and making them more efficient at their job. But would we do this? These decisions would curtail our freedom and flexibility. What happens to someone who chooses not to follow the rules? If you take nothing else from what you've read, take this: the design of working robots cannot be separated from the design of society.

Outfit Our World to Help Robots

In addition to helping robots by organizing our public spaces, we can help them by outfitting our world better for them. That would involve making our signs and other cues clearer for robot consumption.

As we've covered, people are capable of taking in unstructured and sparse information and matching it to patterns they've seen previously. If a street sign is partially blocked from view, we can still piece together partial words and guess what the name of the street is. But robots would struggle to do the same. Also, we can take in many cues simultaneously and pick up on implicit information, such as body language, to gain a deeper understanding of a situation than what could be gleaned via explicitly stated information. Rules of the road can change daily—for example, a new construction sign might be posted—but we can adapt easily. Robots, conversely, take direct measurements of the world and use those measurements precisely, just as they were programmed to use them. This makes it hard for robots to deal with unexpected information or situations and uncertainty in data.

Can we make the world more understandable to robots by adding signposts specifically designed for them, or by adapting the ones we already have slightly, so that they are more machine readable? We have already done this, in some cases, for people who sense differently than the general population—for example, for those who are blind or visually impaired. We've created ways for pedestrians to get critical information without needing perfect sight. A series of beeps begins when a walking person flashes on the pedestrian signal. You can feel the transition from a sidewalk to a street on the bottom of your feet, from the pattern of bumps on the ground (known as a *tactile ground surface indicator*).

We could devise ways of helping robots navigate our unpredictable world more reliably, too, by integrating new perception aids into our environment. Here, we might look to the Roomba for inspiration. Some Roombas come with beacons, which owners can use to create virtual walls that block off spaces they don't want their Roomba to enter. It's like a "Do Not Disturb" sign—and from experience, we can tell you that it works very well when your children have laid a project that they are working on out on the floor. Similarly, robotic lawnmowers often come with barriers to mark the perimeter of your yard, ensuring the robot doesn't veer into your neighbor's yard.

In the future we will need both physical and digital signposts to help robots coexist with us. Sometimes these physical signposts will be new, things added to our world that, just like the magnetic tape in our factories, give robots bread crumbs to follow. We can also modify existing signs and markings in our world to make them more robust for robots—for example, we could add QR codes or other types of signals that robots can sense. We will need to standardize them, make them machine readable, and design them to be impervious to normal wear and tear, so that robots can readily train on them and use them reliably.

The kind of digital information about our public spaces that is already becoming available can also be used to help robots. "Smart cities" are collecting lots of useful information, and they will be able to pass this information on to robots. We already have a rapidly growing database about, for instance, traffic patterns and crime detection. Vehicle-to-everything (V2X), which provides a communication mechanism for vehicles to communicate with infrastructure, networks, other vehicles, devices, and even pedestrians, is slowly rolling out.[12] V2X uses Wi-Fi and cellular communication to enable features as simple as communicating a traffic signal change, or discovering available parking spots and charging stations. Armed with this information, robots would be able to navigate more easily around areas that are otherwise challenging for them.

Adapt with Smart Environments for Smarter Robots

We've already discussed the importance of designing robots with automation affordances so that we can extrapolate or predict their behavior across a wide range of situations. But it doesn't all have to fall on the robot's shoulders. In fact, we can also employ artificial intelligence to design our environments so that both robot and human behavior is more predictable.

Some of these design elements will be adaptations of affordances we already use for people. We have long known that careful design of the environment can effectively guide our own behavior. Exit signs and

LED lighting on the floors direct people along the aisles in a darkened movie theater and are essential in emergencies. Cones and signs on a street direct drivers around roadwork, and barriers and signs at airports might guide you through, rather than around, a duty-free store, so that you are more likely to buy something on your way to your departure gate.

A subfield of artificial intelligence, called *goal recognition design*, develops algorithms that can optimally configure an environment—that is, select where and how to set up barriers and signals—in exactly this way.[13] The aim is to determine how to modify environments such that the behavior of agents is fully predictable as early as possible. A robot that can predict where someone will go is a robot that can maneuver to stay out of that person's way, or plan to collaborate or help in a task. Being able to predict how people will behave as they navigate through an environment or interact with it would be a powerful tool for a robot. Interestingly, even a single carefully placed barrier can make your movement through an open space substantially more predictable. Consider the following scenario: You are entering a grocery store with your shopping list in hand. You can never just make a beeline straight to the first thing on your list. Instead, there is almost always a display to go around. If you take a left, you might be funneled to the produce section, the starting point for the typical "flow" through a grocery store. If you go right, you also start to reveal information about your goals, your priorities, and your likely trajectory. Many different kinds of barriers can create this type of bifurcation in our behavior, some of them in ways that are more intrusive than others. New AI techniques provide an opportunity to automate the design of the environment to provide much-needed structure to reduce midlevel societal entropy.

Similarly, unobtrusive but well-timed signals can modify human behavior at scale. Think about taking a walk across town. Depending on the timing of traffic lights and pedestrian signals, you might cross on one side of the street or another. In your mind, either way is just fine as long as you are making progress toward your destination. Now there's a robot opposite you at the intersection. Depending on whether you cross here or there, either you will delay the robot yielding to pedestrian traffic while it's waiting to make a right, or you will both be able to proceed on your way at the same time. To make this work, the traffic lights and pedestrian signals will have to be adapted to a human-robot world. They will have to be responsive to the very dynamic conditions on the ground. Averages of traffic flow over time will not be sufficient for determining these interactions; instead, the whole city traffic system will have to be organized, in part, around the individual needs and goals of humans and robots sharing space on the streets. Optimizing the trajectories of humans and robots will be reduced to the simple problem of *deconfliction*, just as in air traffic control—but in this case, to prevent and manage conflict city block by city block. With this approach, we'll reduce the confusion in high-entropy areas and make them more manageable.

OVERCOMING SOCIETAL ENTROPY IN OUR FUTURE WORLD

Now think again of the grocery store of the future, where robots follow maps of the store just as we follow Google Maps when driving. Digital markers light up in the aisles to guide robots through the store, and shelves reconfigure themselves to make things easier for robots to reach. The environment comes alive to help robots find their way. Not every store will have the means or the immediate need to do this, but that's all right. The select few that do will provide us with opportunities for experimentation, so that we can find the best ways of reducing societal entropy in those spaces. We'll create space in these early days for users to begin offloading the rote task of picking up groceries to machines. The grocery stores with the resources to do so will be motivated to transform the environment because of the benefits working robots will offer.

The grocery store is a chaotic place, with many things contributing to its societal entropy. Think of the frustration you feel when you're trying to quickly pick up an item on the way home from work, but the item you want is in a new place, or the aisles are crowded with carts of people moving as slow as molasses, some clogging up the aisles as they carefully scan their lists or decide which items to buy. The checkout lines are long. Now imagine a future in which you order your groceries on an app like Instacart, but instead of the order going to a human shopper, it is placed in the queue of a robo-shopper. How challenging will it be for robots to make their way through the tight aisles and find the items you want, all while avoiding interference with bystanders? The store will need to be redesigned to reduce the impact of societal entropy, not only for delivery robots, but also for robots that restock shelves or carry items for customers.

To tackle the problem, we'll have to focus first on what aspects of grocery stores already exist that offer low entropy. Although grocery stores might move individual foods around, for example, their basic layouts—the location of produce, the bakery, the dairy section, the meat department and deli—do not change frequently. It would take too much planning, effort, and money to reorganize those departments, so that doesn't often happen, and societal entropy is low here. When we walk into a familiar store, we already pretty much know where everything is. Our robots will similarly be able to get under way if the store provides them with a digital map of the store layout.

That map would enable each robot to coarsely plan its route through the store, but it would not tell the robot exactly where certain items are, or whether a certain item is missing and out of stock. Inventory can change depending on the date of the last shipment, restocking schedules, and how many people and robots came through looking for a specific item recently. The robot will surely require its own vision and perception system to be able to identify and retrieve items and refine its map as it goes. But how will it plan its route through the store, considering the many high-societal-entropy entities—people, children, and carts bobbing and weaving about?

An understanding of the rhythm of the grocery store's daily and weekly patterns of activity—of medium-level entropy—could make all the difference. For example, say the store doesn't receive a fresh fish shipment on Mondays. As a result, the fish market has a more limited selection at the begining of the week, but there is still frozen fish. Or the robot might know that foot traffic in a certain aisle tends to be lighter at a certain time, so maybe it can zip a little faster by that route to the dairy section. The robot also knows that between 5:00 and 7:00 p.m. there is a wine-tasting kiosk, so it assumes that the alcohol aisle will be quite busy then and can avoid that route. It can coast more easily down sparsely populated aisles, instead of weaving around a crowd that is more interested in the free wine than in anything else around them.

Occasionally, though, the poor robot will have to brave the crowd to retrieve a bottle of wine for its customer during happy hour. There is no avoiding the beverage aisle, and it is simply packed too tight for the robot to pass. What's a robot to do?

First, the grocery store must *organize*. For example, aisles can be designed with a one-way flow of traffic, similar to some city streets. Open spaces off the aisle can be designated for humans who want to pause to sip their wine. Now the robot is guaranteed to make progress down the aisle, even if by following behind a very slow person. A robot crosswalk may be created for robots to safely move through a particular high-traffic area.

Next, the grocery store must *outfit*. Real-time sensing of inventory on the shelves can tell a robot which shelf an item is on, and whether it will need to reach for the last wobbly box of mac and cheese or there is a full row of boxes there. Real-time sensing and reporting of how many people are in different aisles and where can help the robot adjust its route. In this way, robots can follow conventions or guidelines to travel together when possible, creating a flock, and perhaps thereby having less of an impact on customers than if the robots were independently meandering throughout the store. Plus, robots maneuvering down aisles will be much more efficient if they can avoid having to pause for a human who is stopping to scrutinize all the ingredients on the back of a box of waffle mix. By the same token, a robot may plan a path to an open aisle, but see a bunch of people with shopping carts once it gets there. By convention, the robot could give the people space and move on to a different aisle.

Finally, the environment can be *adapted* to direct robots along their optimal routes. Grocery stores are already designed with a flow. You walk into produce, and you are then funneled to meat, fish, and the deli. Next you go up and down the aisles to baking goods, canned goods, and so on, and finally you make your way to the far side of the store, where the dairy products are. Now imagine the grocery store coming alive with illuminated arrows to guide you through your personal, most efficient path through the store, based on current conditions. The shelves themselves are robots on movable platforms, slightly expanding and contracting the size of the aisles and repositioning products depending on the time of day or season. The aisle is as narrow as possible while a flock of robots zips down it, and then expands wider when humans and robots must pass one another. And just as we give extra room and consideration for the employee who is restocking a shelf with a large cart behind her, shelves collapse around a robot that is restocking a shelf, to ensure that customers don't interfere with the robot's task. The grocery store itself becomes a cyborg, reconfiguring to make our hybrid human-robot society possible.

Now, imagine your neighborhood, too, coming alive around you to create an environment for human-robot partnership. Lane lines are lit up with digital markers that change depending on traffic flow. They guide robots into a single flock and use traffic lights to provide spacing between people and robots. The flock splits into two to allow cars to pass when traffic density increases. The robots on the street and sidewalk are brought to a halt as they approach a crosswalk, because the school crossing guard uses a stick to create a virtual wall as schoolchildren skip across the street. Real-time information about traffic patterns and active construction sites allows robots to route around these areas or, if they must enter these areas of high societal entropy, to change their behavior accordingly. A robot breaks off from the flow to pick up a pizza, and then navigates down the sidewalk to deliver it to a nearby apartment. The apartment doorways are lit up with digital markers, so the robot can find the exact unit for its delivery. The city also is a cyborg, helping to guide robots so they can manage the wide range of societal entropy that they constantly encounter.

CHAPTER NINE

It Takes a Village
to Raise a Robot

IT WAS MAY 2016 IN FLORIDA WHEN A DRIVER SET THE CRUISE control to seventy-four miles per hour in his Tesla and turned on autopilot.[1] The car was in control—time to stretch out and relax. The driver had come to trust the autopilot and rarely put his hands on the wheel when it was on, despite the warning message that came on every time. The message told him that autopilot was "an assist feature that requires you to keep your hands on the steering wheel at all times." But he only had his hands on the wheel for twenty-five seconds of the thirty-seven-minute drive.

Thirty-six minutes in, a high-height tractor trailer crossed the highway in front of the car and its sensors didn't identify it. The white side of the truck was hard for the car's camera to see against the brightly lit sky. The unusual height made it look like a roadway sign to the car's radar.[2] At full speed, the Tesla plowed into the truck. Its roof struck the bottom of the tractor trailer and it was torn off. Even though the driver was hunched over unconscious, the car passed under the truck and continued to drive at seventy-four miles per hour eastward on the highway for about a mile. It left the roadway, smashed through two fences, and struck a power pole, at which point it finally came to a stop before doing any more damage.

You may think this was a naive or lazy driver, unaware of the potential pitfalls of the technology driving his car. But he was well aware of the dangers of autonomous driving, considering some comments he had posted on social media about Tesla:

> Bigger danger at this stage of the development is getting someone too comfortable. You really do need to be paying attention at this point. This is early in the development and the human should be ready to intervene if [the autopilot] can't do something. I talked in one of the other comments about the blind spots of the current hardware. There are some situations it doesn't do well in which is okay. It's not an autonomous car and they are learning HUGE amounts of data about the car doing the driving. I'm happy to help train it. I'm VERY curious what version 2 of the hardware will be like and what [it] will enable.[3]

If even this driver became complacent, when he knew he was an active participant in training and improving the technology in the face of blind spots, what can we expect from members of the general public, who will often find themselves, willingly or not, speaking with robots that are still learning how to act?

Tesla reported that this is the first known fatality in over 130 million miles where autopilot was activated. This is on par with the overall automotive fatality statistic of 1.25 deaths per 100 million vehicle-miles.

But how many near-misses or incidents occurred prior to this tragic accident? In publicly reported statistics, drivers take over from automated driving capabilities to address a concerning situation at a rate of once every five thousand miles.[4] What can we learn about these vulnerable moments to address them before they cost us more human lives?

In the case of the Florida accident, Tesla claims it has addressed the problem. But how exactly it did so remains a mystery. What about other autonomous technology and car manufacturers? Are they also able to learn from this tragic event and prevent such a scenario with their systems?

We don't know the answers to those questions, because currently robot manufacturers hoard their data. A widely quoted reality in the expanding field of AI is that "more data usually beats better algorithms." The community acknowledges that the performance of a system greatly improves as more data is gathered and used to train it. But that also means there is an intrinsic market incentive to tightly protect data as systems are deployed. What manufacturers haven't fully realized—or don't yet want to admit—is that this isn't only about keeping company data proprietary so you can dominate the market. This is a matter of whether robots are as safe as they can be. It's a matter of life and death.

Data is valuable, because the rules employed by algorithms are designed by people and hand-tuned to make a product capable of dealing with a wide range of scenarios. And because it's so expensive to develop, it's no wonder that companies want to keep this proprietary information private—it gives them an edge over the competition. Despite the many successes we've seen in the tech sector in recent years, however, this is a fundamentally flawed way of proceeding, because people aren't capable of imagining all the situations that an algorithm may have to make decisions about. In machine learning, on the other hand, the robot or system learns the features and patterns from the data itself. The more data you collect and train it on, the more scenarios the system will be able to handle. We therefore owe it to society to share data and elevate the performance of all the robotic systems we introduce into our everyday lives. The development, tuning, and refinement of robotics depend on access to knowledge about how automation is used and how it performs in real-world scenarios. Data about what happens when a robot encounters situations the designers never expected to occur is especially crucial for making systems better. Manufacturer testing usually covers all the known pitfalls of a new technology and a subset of known challenging situations. It's the unknown interactions that go untested. By definition, they are at first unknown, waiting to be discovered as a robot enters the real world. If we had centralized collection and sharing of data about incidents, accidents, and safety concerns, we could fast-track this learning process around the world. Indeed, that's likely the only way to do so.

How long will the public stand for robotics manufacturers hoarding their data in secret as the number of preventable accidents mounts? The only way to make robots that can survive in the real world is to learn from the mistakes we make in designing them—and to share our hard-earned wisdom with each other.

IS THIS A REALISTIC EXPECTATION?

You might think it's unreasonable to ask companies to give away their proprietary assets and it would hurt competition. But do you know what hurts industries more? A catastrophic incident that ruins public trust.

We can look again to the aviation community, one of a few industries that work just fine sharing data at scale across manufacturers, airlines, and unions. The creation of a system to collect, aggregate, and analyze operational data was motivated in part by testimony from William Patterson, the president of United Airlines, before the US Senate in 1958.[5] Patterson was testifying in hearings about the proposed Federal Aviation Act, which created the FAA. The act unified the control of airspace and changed the way safety rules were created and enforced. During his testimony, Patterson called out the importance of sharing safety trend data across the industry. He pointed out that data about incidents that fall short of an accident can give us great insights about operational problems and allow us to correct these problems before they result in tragedy. In his words, by sharing safety trend data, we can "act before the happening." Bobbie R. Allen, the director of the Bureau of Safety of the US Civil Aeronautics Board, later referred to the accumulated statistics on minor aviation safety incidents as the "sleeping giant."[6] As a result of these efforts, pilots and air traffic controllers now submit reports on safety incidents even if they do not result in an accident. These can include a malfunctioning navigational aid, a breakdown in procedures between the pilots and air traffic control, or any other safety-related event. An average of two thousand reports are submitted weekly and over sixteen hundred queries of the database are performed every month.[7]

There are many corollaries to what we can expect as robots enter our everyday lives. Every type of robot, every manufacturer, and every user will experience minor safety incidents, and each one will contain some nugget of insight. If we aggregate these, we will be more likely to anticipate and prevent larger problems. If that data remains hidden, we will learn less, it will take longer to learn, and it's more likely that what we do learn will come only after some catastrophic events occur.

So how did the aviation industry manage to get people and companies to share this valuable information? The federal Aviation Safety Reporting System (ASRS) was established in 1976 and is managed by NASA through an agreement with the FAA.[8] Under its auspices, people involved in aviation make confidential incident reports on a voluntary basis. NASA, as an independent entity, protects the privacy of those reporting information, maintains the integrity of the data, and prevents anyone reporting such information from suffering any punitive action.

The FAA outlined the purpose of ASRS in a memo to NASA in 1979: "This system will be designed primarily to provide information to the FAA and the aviation community to assist the FAA in reaching its goal of eliminating unsafe conditions and preventing avoidable accidents."[9] Its objectives are:

- To create a confidential reporting system that any person in the national aviation system can contribute to
- To operate the digital system to store and retrieve the reported data
- To provide a system to analyze and study the gathered data
- To create a responsive system to share the results with those responsible for aviation safety

Pilots and air traffic controllers can log in personally and report any incident, even their own errors. Incidents range from minor failures or errors that have no consequences to near-misses that were seconds away from a disaster. The voluntary reporting of information directly from the source results in higher data quality than pure interviews or other mandatory reporting structures. ASRS is now over forty years old and has taken in 1.5 million reports. It has put out over 6,000 safety alerts to the aviation community. Figure 21 details the subjects of alerts issued just in 2017. These incident reports influence how the industry looks for structural problems.

For example, consider the issue of wake vortices. A wake vortex is a disturbance of the atmosphere created by an aircraft. It is highly affected by runway positioning and the spacing of aircraft. Turbulence from wake vortices is a particularly big problem for takeoff and landing, and many reports have been submitted to ASRS.[10] One report, from Phoenix, Arizona, for example, said that a "B737-800 Captain reported encountering 'severe' wake turbulence departing PHX in trail of an A321."[11] The report from Phoenix was only one of fifty summarized in a subsequent report issued by the ASRS looking at incidents involving wake turbulence. The ASRS analysts evaluated the reported magnitude of wake encounter, aircraft spacing, aircraft type, runway configuration,

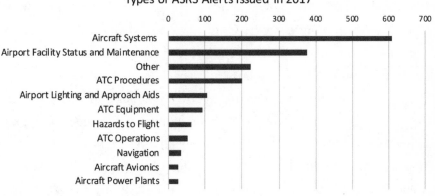

FIGURE 21: A summary of the subjects of ASRS alerts issued in 2017.

and consequences from the encounters with wake vortices to better understand factors affecting the frequency and magnitude of turbulence due to wake vortex and adjust procedures and spacing accordingly. In a future world, if we have a reporting system for robots, we may see a similar study about the interaction of robots around challenging intersections to help in improving the design of the intersection and the logic of the robots.

The model for improving technology, procedures, training, and operation through a centralized reporting system has been proven in the aviation community. ASRS allows us to wake the sleeping giant and see into the cockpit across millions of scenarios. We can see how pilots and air traffic controllers operate in the face of equipment malfunctions or environmental anomalies. These diverse and challenging real-world scenarios are things we could never uncover in engineering or qualification tests at this scale. And then we can make things better by improving the cockpit or the aircraft spacing in certain areas or adding new training.

Without a doubt, ASRS has increased the safety of our skies. As a result of the success of ASRS, there are now aviation safety reporting systems around the world, including in countries as distant as Brazil and

Singapore. Even the Federal Railroad Administration (FRA) adopted the model and created its own reporting system, called the Confidential Close Call Reporting System (C3RS).[12] What should such a system look like for robotics in order to capture and learn from the millions of interactions and anomalies robots will face?

But even if we can agree that tech companies should be willing to share data, you may wonder if this amounts to treating a symptom rather than a cause. Surely, we should be able to catch these problems during a product's development and testing process. After all, engineers spend thousands of hours carefully designing systems, analyzing how they may fail, and testing them before putting them into operation. Although that argument may sound compelling enough, it is naive. It is widely accepted in engineering that there are behaviors that occur in complex systems that take time to prove out and understand.

Systems as complex as aircraft and robots have many thousands of individual components. Each of these is *deterministic*, meaning that it should behave the same way every time. When you flip a given switch in the cockpit, it should always send the same signal to the control box. But when you put all of these components together, and then place them in a dynamic world, you lose the ability to predict how the system will behave in every situation. There are just too many configurations of those deterministic components. This is known as *emergent behavior*.

More precisely, as defined by technology historian George Dyson, emergent behavior is "that which cannot be predicted through analysis at any level simpler than that of the system as a whole." Moreover, Dyson said, "explanations of emergence, like simplifications of complexity, are inherently illusory and can only be achieved by sleight of hand. This does not mean that emergence is not real. Emergent behavior, by definition, is what's left after everything else has been explained."[13]

Jeffrey Mogul, working at HP Labs at the time, referred to the example of London's steel suspension Millennium Footbridge, which has come to be known as the "Wobbly Bridge."[14] It was designed by civil

Figure 22: The Millennium Footbridge.

engineers and went through all of the typical engineering analysis and testing to make sure it was safe for people to cross. However, once it opened, it had "unexpected excessive lateral vibrations" on its opening day that were only discovered once many people started crossing it. The bridge started swaying as pedestrians walked across, and the pedestrians responded by compensating with their footfalls, as we do on a swaying train to keep our balance. All the pedestrians fell into this pattern, stepping in synchronization, and it caused excitation, as when we use our legs to pump a swing. Except the designers had not anticipated this synchronization, and the wobbles became more exaggerated with every synchronized step. On opening day, more than eighty thousand people crossed the bridge, with two thousand on it at a time. As the wobbles worsened, it was decided to close the bridge to determine what the problem was, and it took nearly two years to design a fix. In the end, the bridge was modified with dampeners to eliminate the motion.

If engineers can fail to predict the emergent behavior of a bridge, a technology that has remained pretty much the same for thousands of years, how can they possibly predict the behavior and anticipate all the problems that may emerge with intelligent robots that have to survive in our chaotic world? Well, quite frankly, they can't. Emergent behavior can only be discovered by testing the full system in a realistic environment. This means that some degree of failure of everyday robotic systems is fundamentally unavoidable. So our goal should always be to devise ways of trying to learn about a system's emergent behavior before it is introduced to the public.

One method that has proven popular is simulation. Engineers build virtual worlds and virtual representations of the types of environments a system will experience so that all or part of the integrated system can be tested. For example, before a spacecraft is ever launched, the entire system goes through layers of virtual testing: parts of it are even submerged into a giant pool to mimic the weightlessness the system will experience outside Earth's atmosphere. The virtual worlds allow a system, or at least large parts of a system, to experience thousands or even millions of different situations before it is fielded. The results of this experimentation are then used to identify failure points and correct them.

The industrial world develops multiple virtual representations of complex systems to test different aspects of their design throughout the development process. Early in the development phase of a machine, engineers build what is referred to as software-in-the-loop (SIL) simulation. This happens before any hardware is even made. Engineers build a software representation of the system that can be executed in a virtual world, like a video game. Then, once real hardware is developed, they test it again with a hardware-in-the-loop (HIL) simulation. The parts of the fully virtual system that represent the new hardware are swapped out for the physical version.

Say, for example, you constructed a number of virtual scenarios in which a robot passed people on the sidewalk. At first you would test your sidewalk delivery robot with a robot simulated in software, along with simulations of various types of people approaching, as well as a simulated perception system. In this example, let's say the simulated perception system mimicked (in software) the light emitted from a pulsed laser system and bounced off objects in the environment. Once the physical robot became available, you could deploy it to navigate a corridor in an office building, but instead of using physical people, you would test the hardware in conjunction with a simulation of people, including a simulation of the LiDAR sensing people approaching. Hardware-in-the-loop simulation allows the engineers to plug in real sensors, real embedded computers, or real actuators to see how all of these play together in the virtual world. Finally, once the entire system is developed, they would use many physical virtual worlds for full-scale integration testing or qualification testing. For example, the US Army maintains Yuma Proving Ground in Arizona as a wide-open space for the testing of many complex systems, including weapons and autonomous systems. These types of testing areas typically feature different environments that can re-create many realistic scenarios. Those charged with carrying out the tests collect detailed data during the testing itself and learn from the results.

Finally, there is human-in-the-loop (HITL) testing. One theme we have revisited repeatedly throughout this book is that, to some extent, designing robots is hard, because people are inherently unpredictable. It's important to find and correct failure points by testing with operationally relevant people. All three layers come together in HITL testing: subjects can play a video game, interface with real hardware, or serve as actors in the physical test environments.

As we look to apply these approaches to developing working robots we face a new challenge. These approaches are costly to develop and maintain, and a consumer product company may not have the same resources as an aerospace company. For example, it is widely accepted in the aerospace community that the amount of effort needed to test a piece of software is equal to the amount of effort needed to design and develop it.[15] In other words, for every software developer you have, you need another person dedicated to testing and trying to break that software. The resources required to certify safety-critical machines pose a formidable barrier for consumer companies. First, we would need a set of shared resources to model and virtually test individual robot capabilities across the layers described above. And we'll have to use real data from operations to refine this virtual world, because while these models help to cover a large number of scenarios, they cannot cover every possible situation a robot may encounter.

There is already some precedent for this in the robotics community. Robots employing deep-learning techniques might require millions of trials to learn how to grasp everyday objects.[16] However, this data can be collected quite efficiently in virtual worlds that import computer models for objects, simulate the data produced by a robot's depth-image sensor, and test different grasp strategies with a physics simulator. In a strange and exciting new line of work, these virtual worlds can introduce randomization to create situations that would never plausibly be encountered in the real world. In other words, the robots are trained on artificially created simulated worlds that look like nonsense to a human eye. When we insert a wide range of random noise, such as blacking out pixels, blurring them, or adding random objects, real robots are trained to be more robust to the randomness we see in the real world. The robots are able to identify objects and select grasp points in the more predictable real environments. The artificial training data sets can be used as a kind of stress test, to train for situations that are actually more challenging than those encountered in the real world.[17]

This is a good start, but we need to go further in order to successfully design, develop, test, and qualify complex systems of interacting intelligent robots and people. We need to rethink how we gather and use data, because the characteristics of the data used in machine learning effectively determine the robot's behavior. Component-level analysis, simulation, and testing must now address a whole new set of unprecedented challenges as machine-learning systems interact, learn from, and influence people. That is, unless we are comfortable with our everyday streets, sidewalks, and stores serving as our "Yuma proving ground."

QUALITY OVER QUANTITY

Consider, in the not-too-distant future, that a sidewalk delivery robot is designed, tested, and deployed on the hilly streets of San Francisco, and then is shipped across the country to Miami for its first East Coast deployment. New York and Boston were considered as acceptable second cities, but Miami was selected as an ideal location because of its wide sidewalks and temperate weather. The robot was "hardened" to function well with the corrosive, salty sea breeze, but was not designed to contend with snow and ice.

In San Francisco, after a few initial, harmless mishaps, the robot learned to zip around young professionals on the go and maneuver cautiously around baby strollers. But once it's transplanted to Miami, the company notices the system is functioning far below its target metrics. It takes twice as long to get to its destination, and it frequently halts, through use of its speech-command feature, whenever a bystander says "Stop" or "No." The average age difference in the demographics of the two cities proves to be a problem. Miami's sidewalks have many more retirees, and they walk slowly, some using canes or walkers. In San Francisco, the robot learned through bystander feedback to give wide berth to any slow-moving pedestrian. But in Miami, many more people walk slowly than in San Francisco. After a few encounters in which the robot's safety stop was triggered, the robot quickly updated its behavior to act even more conservatively. The training data gathered in San Francisco was too different from the situation in Miami, and in an effort to learn and compensate, the robot's intelligence literally stopped it in its tracks. The robot company's business model has been upended, and further rollouts are halted.

It could have been worse: the robots, unprepared to deal with elderly humans, might have collided with them instead of stopping. So this was the best-case scenario, given the situation. And unfortunately it illustrates that next-generation intelligent robots will face new challenges beyond those encountered in aviation and the engineering of other complex systems. Cockpit automation is a "frozen" system. It is designed, certified, and tested for years before it makes its way onto an airplane. We collect data from the ASRS, analyze it, and issue alerts to take action to make our airways safer. And unlike our new robots, cockpit automation doesn't incorporate modern machine-learning technologies. Robot behavior will continually evolve, driven by data incidentally collected. This is in contrast to aviation, where the system evolves through purposeful, careful analysis and decision-making of experts (like those analyzing ASRS) who are trained to consider engineering, procedural, and organizational factors. What's more, airspace is governed by very similar rules worldwide, whereas social rules are often not global and are in constant flux. Although our new working robots will benefit from data-driven machine learning, to adapt better to our environment over time the deliberate rollout of these updates is of equal importance to advancing robot capabilities.

We hear the same chorus of machine-learning experts and non-experts alike—new intelligent robots will be so intelligent, so capable of learning on their own, that they will quickly exceed the performance of human-engineered systems. The claim is that new, more intelligent machines will make the old paradigm of experts analyzing data obsolete. This couldn't be further from the truth, especially for systems of systems. Even at the single robot/component level—independent of the human interaction—the cracks in this argument are already showing.

There is a machine-learning adage that says "More data beats better algorithms" (i.e., better engineering). It dates back to a 2001 study conducted at Microsoft Research.[18] In analyzing the performance of a variety of speech-recognition algorithms, researchers found that the amount of data used to train the system was a primary factor in performance. Studies of other machine-learning systems found similar trends, and a decade later Google's research director, Peter Norvig, famously claimed, "We don't have better algorithms. We just have more data."

But over the years since, researchers have uncovered a more complex story.[19] More data clearly helps with machine-learning systems where many parameters must be precisely tuned—as is the case with speech-recognition systems. These more complex models are not always helpful, however. The models only learn what is in the data set, and if the model becomes optimized for only the scenarios in the data set, then they won't work when the real environment deviates even slightly from the training environment. In our example of the robot trained on the sidewalks in San Francisco, the robot learned an interaction model that did not transfer well to the different demographics in Miami. This problem is called *overfitting*, and it occurs when you produce a model that so closely replicates the particulars of a data set that it cannot make useful observations about any other similar set of data. To avoid overfitting, often machine-learning experts design simpler models with fewer parameters, so that the machine can learn more general behaviors that may transfer better to new situations.

In these cases, with simpler models, more data may not help—at some point the parameters are tuned as well as they can be. Rather, it is higher-quality data that makes the difference. The data must be engineered. If the data is "cleaned" to remove extraneous factors and outliers, the machine-learning model can learn more efficiently. For example, the robot in San Francisco learned to maneuver carefully around baby strollers but not the elderly. However, a machine-learning expert might note that people pushing baby strollers typically move slowly, as do some elderly people with canes, walkers, or even wheelchairs. If the machine-learning expert had withheld the data that visually identified and labeled strollers during training, the robot may have learned a more general model to approach and pass any slow-moving person cautiously. If the robot had approached the elderly cautiously in Miami in its first interactions, it would have received fewer safety-stop commands, and would not have updated its behavior to act even more conservatively. Problem solved? In this instance, possibly. The careful cultivation (i.e., engineering) of the training data might have successfully addressed this one problem.

But hindsight will always be twenty-twenty. And every high schooler is cautioned about removing outliers to make their data look better. We can easily inject bias into the data, which will manifest itself later as a hidden bias in the machine-learning model. In 2018, in a study from the MIT Media Laboratory and Stanford University, researchers found that a set of commercial face-recognition systems—from Google, Microsoft, and IBM—were quite accurate in predicting the gender of people with light skin tones, but made mistakes up to 20 or 34 percent of the time for women with dark skin tones.[20] Presumably, the underlying data sets were skewed and included fewer women with the darker skin tones. But the researchers couldn't know for sure that this was what happened, because the training data sets were not made available. The disparities were quickly corrected by the companies after the release of the article. But without information about the characteristics of the underlying data set, we are left wondering what other surprises may crop up down the line—in this system and in other commercial AI technologies.[21]

In other words, *better data beats more data*, and you can't guarantee your data is good if you never let anyone else look at it. Even for simple, single machine-learning systems like face-recognition systems on intelligent robots, we must have a mechanism for sharing of data—its properties and characteristics—in order to work together to ensure that these new systems are equitable.

Improving the quality of our data is crucial, but possibly the greatest challenge to collaborative robotics, still below our collective consciousness, for the most part, is how we will grapple with the complexity that arises from deploying these intelligent systems at scale throughout society. A robot may work well in San Francisco or Miami, but surely other cities would offer other challenges. Remember, the whole reason those engineers picked Miami over Boston or New York was that the robot wasn't capable of dealing with snow and ice. And we haven't even talked about varying cultural norms, such as which side of the sidewalk to use to pass other pedestrians, or how close to get to a person when you want to converse with them.[22] No single company will be able to foresee all the new emergent interactions and behaviors working robots will face. Meeting these challenges will require collective efforts, on the scale of the ASRS and the Yuma Proving Ground. We may need a Robot Safety Reporting System (RSRS) as well as human-centered test environments—constructed populated towns and cities akin to the Yuma Proving Ground—to capture, share, and learn from the near-misses, errors, and social benefits and frictions of the gaggles of intelligent robots on our doorsteps.

A PATH FORWARD

The good news is we are starting to see the first signs of large-scale sharing of data for use across companies and research institutions. ImageNet is one example—a database of visual images with objects labeled within the images.[23] As autonomous cars become more prevalent, companies are beginning to release labeled data sets logged from car sensors in challenging situations like rainy days or busy city intersections. There is also increased effort to cultivate full, realistic data sets, including about bystanders, as interactions become more prevalent. For example, Aptiv's nuScenes has a database of over a million images, and over a million people are labeled within them by their characteristics, such as "human pedestrian construction worker," or "human pedestrian child."[24]

These kinds of databases are treasure troves for the research community, but there is more to do to seamlessly integrate robots into the fabric of our society. These labeled data sets cover one dimension of the type of data needed: they help engineers train a robot's perception system. The training will only help a robot better identify a person or a tree branch. It will not tell the robot what to do with that information.

Importantly, these labeled data sets are also missing information about when the overall system fails. It is up to the researcher to discover these problems and analyze why failures occur and how to overcome these vulnerabilities for each individual system. We want robots to be as smart as possible, but we also need to know how they behave when there is an error in perception or other steps in their processing. For example, if a robot confuses a person with a tree branch, that may be okay, as long as the robot is supposed to treat a human and a tree branch the same way. It may be okay to bump a tree branch, but not a person, however, so if there is uncertainty as to whether the robot may think a person is a tree branch in some situation, then it shouldn't bump the tree branch without first resolving the uncertainty to an acceptable level. Perhaps it should take steps to mitigate potential harms—for example, by alerting its supervisor or a bystander.

There are many ways we can expand our support for the robotics community to better design, build, and refine robots, not only while they are in the field and experiencing unexpected situations, but also during the design process. We can provide data, models, facilities, and processes to give the developers the best shot at discovering emergent behavior throughout the development cycle. At present, companies primarily rely on their own resources to design, develop, and test robots. But there should be ways for the greater community to contribute. The greater community could provide more data than any one company could ever create or have access to on its own, raising the chances of success for all.

During the design phase, these labeled data sets are critical, because machine-learning systems are designed by the data. Where there are standardized data sets, we've seen leaps and bounds in improvement in the technology. ImageNet, launched in 2007, has over ten million images and identifies twenty thousand categories of objects in them, ranging from "balloon" to specific dog breeds. Because of this publicly available data set, we've seen the error rate of correctly classifying an object in an image go from 25 percent down to 5 percent or even better.[25] Developing and maintaining labeled data sets that extend beyond object classification and include acceptable or unacceptable robot behaviors or interactions needs to be a systematic offering to the greater robotics community that is expected rather than rare.

During development, if there are virtual worlds available, then engineers can try out the various components of the system or a virtual representation of it long before the entire thing is built. Scenarios can be modeled and made available to represent challenging use cases that robots will need to handle. This is still early enough in the process that developers can make significant changes to the robot based on what they find.

During the testing process, the manufacturer tests the entire end-to-end system for the first time. This is the first opportunity to discover emergent behaviors. Many manufacturers have their own proving grounds or physical test environments, such as racetracks or gravel surfaces and stairs that lead to nowhere. But not all companies can afford such infrastructure. In some industrial applications, there are shared facilities for testing prototype systems in very realistic settings. Pooling resources allows for more sophisticated test infrastructure and shared learning about testing protocols—as with military proving grounds like Yuma. For robots, we may need a test city of the future to put robots fully through the paces before fielding them in our real cities and neighborhoods.

Today there is no formal qualification process for robots as there is for many other sophisticated systems, such as aircraft and automobiles. In those industries, this usually includes quality standards with pre-scribed tests and procedures before the systems are fielded. There are some electronics and other hardware standards that apply to any com-plex system, but no such thing exists for intelligent systems. Hardware usually has to function across a temperature and humidity range and meet electrical standards. But what does this look like for intelligence? How should the robot's brain and its ability to partner with the people around it be tested? Under what conditions is it okay for robots to fail? These criteria need to be defined, validated, and maintained so that all robot designers and manufacturers are held to the same thoughtfully designed quality standards.

Once these systems enter the real world, they will inevitably still have much to learn. There will be emergent behavior that hasn't been predicted, situations when the robots will fail. This is when we need a robot equivalent to ASRS so that all robot developers can learn from each of these mistakes, especially "the sleeping giant" data sets, so that we have a chance to minimize the number of catastrophic failures across the industry. As with ASRS, this will only be possible if there is an anonymous reporting system and a process for analyzing the data and then distributing results and recommendations to the broader com-munity. The recommendations may include design changes, infrastruc-ture or environment changes, training modifications, or other ideas to mitigate the discovered problems.

ENVISIONING AN RSRS OF THE FUTURE

The rationale for an RSRS is the same as the one for the ASRS: to share our collective experiences and challenges, to find trends that provide early indicators of hazards, and to correct them "before the happen-ing." The ASRS is administered by an independent body, NASA, and processes nearly one hundred thousand reports a year, with an annual operating budget on the order of millions (as of 2005).[26] Meanwhile in the United States we drive over five billion miles a day. In a near-term world when all of our cars have some automated driving capability, if there were one automated incident report submitted to the RSRS for every five thousand miles of driving, there would be one million reports submitted daily. And that's only for autonomous cars—now add to that the reports from millions of working robots in our homes and offices and on our streets and sidewalks. Processing these reports for trends would be an unsurmountable task. Rather than reports carefully crafted by trained pilots to describe the key aspects of an incident, each of these automated reports would basically be a "data dump" that would have to be excavated. A data dump about a car's sensors, however, may not capture the nuance of the situation: What if the car failed to behave properly because of something it didn't sense? We might envision augmenting the data dump with a "bug report" from the human supervisor or operator, or from the bystander who had the negative experience with the robot. But these people do not usually have an expert grasp of the automation, and may not really know what wasn't working. Finally, one car's sensors and decision-making process might be so fundamentally different from another manufacturer's that the raw data would not provide sufficient information. For formulating new tests for the system to pass, and ensuring safe operation in a newly encountered situation, more nuanced information will be needed.

All this is to say that our solution in aviation cannot be transferred exactly. An independent body may be necessary for anonymity and protection in reporting—which is crucial to the effectiveness of ASRS and other systems like it. However, manufacturers themselves must absolutely be collaborators in figuring out what data is useful to share in general, interpreting specific data collected—which may be inextricably tied to the proprietary aspects of their engineered system—and making an interpretation and test suite available that others can use in development, testing, and improvement. New structures may need to be developed to incentivize this type of public-private partnership. Companies must be rewarded for exposing a discovered flaw in their system and communicating the problematic situation in a manner that can be reproduced by another company.

There is an analogy here to "bug hunts" in the commercial space, where individuals or groups are rewarded monetarily for finding flaws in a system. Similarly, imagine a robot company discovering a problematic situation for its robot. The engineers go to work to understand the root problem—say, its autonomous planner gets confused on windy paths with oncoming pedestrians. They develop a new scenario to represent this challenge in the virtual world—not linked to their own robot's data or sensor feed—and upload it to a "bug-hunt" database. They demonstrate that their robot now passes the test flawlessly. Now other companies rush to demonstrate that their systems, too, perform well in this new scenario in the virtual world. Each system is ranked according to a publicly reported index on its ability to handle the bugs in the database. Think of it as a J. D. Power and Associates ranking of sorts. J. D. Power is a data analytics firm that conducts surveys in a number of different industries on matters such as consumer satisfaction. But JDPOWER could alternately stand for "Jointly Developed Program of Warnings, Engineering, and Robustness."

ENVISIONING A PROVING GROUND FOR WORKING ROBOTS

The Yuma Proving Ground is administered by a central authority, the US Army, and it is used for the testing of new technologies with the simple goal of ensuring that they work as expected for military personnel and as required in near-real environments. The army is the buyer of the technology, and the goals for the technology are clear—to benefit the army. Trying the systems out at Yuma is a matter of proving the technology will work and solving technological or operational challenges before they arise in the field. More than three thousand people, mostly civilians, work at the proving ground, and significant testing infrastructure is available. In addition to technology testing, another core purpose of the proving ground is training military units in realistic environments prior to their deployment overseas. Dozens of units rotate through each year, getting ready for their deployments in the same environment as some of the most advanced tech on Earth. The army, as the central authority, plans the tests, coordinates the use of shared resources, and makes decisions when conflicts arise, such as radio frequency interference arising from concurrent testing of new technologies.

There would be tremendous social value in a proving ground where we could test working robots. In this "Community of Tomorrow," developers could share and learn from near-misses and errors, explore social benefits and frictions, experiment with new infrastructure and organizational changes, and learn how best to evolve our social norms. As at Yuma, the test community would give us a way to test and confirm the potential benefits of intelligent robots and anticipate unintended consequences before the technology was deployed in particular communities or at scale. The Community of Tomorrow may offer real-world, virtual-simulation, and mixed-reality elements, so that we can test technologies in different phases of development. We could then leverage the data and insights collected in the real-world Community of Tomorrow to develop virtual worlds that companies could use in testing and development.

For the physical, real-world component of the Community of Tomorrow, we are talking about an investment on the scale of Yuma, built with functioning infrastructure just as you would see in a town or city. But which town or city? San Francisco, Miami, or a small town in the heartland? The Community of Tomorrow will have to be planned with intentionality to ensure that testing addresses the needs of different groups as thoroughly as possible. Much like the military units who rotate through Yuma to practice and simulate their lives in the field, we would also need residents living in the Community of Tomorrow who would intersect with the technology as they went about their activities of daily living. Who would these residents be, and where would they come from? An opt-in model might result in a selection bias for those who are predisposed to like technology. Offering compensation or other benefits, such as housing, may encourage broad participation and the ability to draw together the right participants for particular technologies or tests. Depending on the particular structures for participation, we would have to include representation by disadvantaged groups. Only by ensuring their participation could we address any potential issues that might cause new technologies to place undue burdens on them.

A facility of this scale is beyond what any single company—even a mega-company—can undertake. It would require societal investment, the same as the FAA. However, unlike at Yuma, a central authority could only succeed in administering a Community of Tomorrow with substantial, deep cooperation of the industries developing and deploying the tech. In the case of working robots, the US government, as well as governments from other countries, are invested stakeholders, but not buyers. Moreover, working robots are a safety-critical, *consumer* technology. Each company bringing technology to test in the Community of Tomorrow is incentivized to sell more products to their consumer buyers. The tests that a particular company may prioritize, therefore—for example, to understand people's emotional response to a new technology—may conflict with the tests required to understand its impacts and ensure its seamless integration into society. A particular company may be primarily concerned with ensuring that the Community of Tomorrow's participants tell their friends about the product and publicizing positive interactions with the robots, and thus disinclined to offer up their most difficult situations to testing. This would be out of concern that participants would reject the technology, or that engineers from other companies might observe their failures and gain a competitive advantage from the information.

Yet, as with the RSRS, in the Community of Tomorrow the engineers and experts who developed the technology would be the ones in the best position to understand the root cause of a particularly challenging situation and develop the test situations that best exercise the potential deficiency. It's not clear that a centrally managed government authority is the right structure, but surely an independent organization will be required to negotiate the many competing goals and priorities of participating companies, critically revise and review experiment protocols, maintain integrity of the process, and ensure that working robots will work for us all.

FUTURE OUTCOME

Now think back to the company that deployed robots in Miami. With this future vision, it would have gone through the following process to prepare for the rollout:

Design: The company would use community-wide data sets to prepare the robot to enter the Miami environment. These data sets would allow the engineers to refine the intelligence of their robot and tailor it to Miami-specific scenarios. Once the robot was trained on the generalized data set to create baseline behavior, the engineers would turn to more specific data sets representing activities and scenarios typical to Miami to calibrate and tune the behavior of the system. For this case, an urban environment with a larger population of retirees would be represented in the more specific training data set.

Develop: During the development phase, the company would tune the parameters of the virtual world to match the setting in Miami. They would have a wider range of variability in driving and walking behavior and a higher percentage of people in wheelchairs and with canes or walkers. The software would be continuously tested in this representative virtual world and be put through the paces of the challenges that are more frequently present in Miami.

Test: Before deploying in Miami, this newly calibrated and trained robot would go to the Community of Tomorrow test facility and operate on real sidewalks, interface with actors behaving like people in Miami, and encounter traffic lights and active construction zones. The engineering team would learn a tremendous amount by observing the emergent behavior of the robot at this proving ground and have the opportunity to tweak the robot to adjust its behavior based on what they learned.

Qualification: Standards would be used to make sure that all test cases were covered and that the design of the intelligent platform conformed to standards that the general population expects for a system that may pose a threat to their everyday lives. Shortcuts in the design and testing would not be tolerated, and the performance of the intelligence would be proven and documented before it entered the world.

Fielding: After fielding, the company and those operating the robots would report any problems that arose to RSRS, even if they were not the fault of the robot. The problematic situation would be posted on JDPOWER, and other robot companies would strive to demonstrate their capability to address this identified problem. As new solutions were developed, a change to the robots would be released, and all robots in the field would be upgraded to prevent more serious problems.

The outcome of unleashing robots in a new city like Miami would be different if we joined together to provide support for creating robots that could help us in our lives. Working together, we have the chance to realize the huge potential benefits that robots could offer to society (saving lives, reducing congestion, providing better access to the resources needed by all). And we could do this without the potential pitfalls that we have already begun to see, including new safety risks and robots creating a nuisance on our streets, on our sidewalks, and in our hallways. But without these shared resources, we will only see incremental improvements in our data and approaches, followed by setbacks, as each individual company attempts to cover the high cost of the development cycle alone. Any one company is unable to achieve this kind of reporting and testing quality. We need a village of policy makers, local governments, and community-wide consortiums to step up and fill the void that exists today.

Conclusion

JULIE RECENTLY WENT TO A DINNER, AND THE TOPIC OF THIS book came up. Her host told her she had bought her mother a Tesla. And her mother loves it! She uses the autopilot feature every day. But her host wondered . . . In the morning when she starts up her own Tesla, if there is a software update she puzzles at the long, convoluted description of it. She's in a rush getting her kids where they are going and quickly clicks through it. We all get several such notifications every week about the technology we own. If you're like us, you probably only glance at them. They're long and jargon-filled, it's often unclear what they actually mean, and they almost always arrive at inopportune times. Julie's host was sure her mother didn't understand how the Tesla might change with each new update, and she wondered what it might mean for her mother's ability to operate the car safely. She was right to be concerned.

On our streets and sidewalks, in our hospitals, apartment complexes, and grocery stores, we are encountering the first wave of working robots, and we pause to consider them. We can imagine the conveniences we might enjoy, or the time we'd get back to spend with our children, as robots become more advanced. For some, these robots can be life changing: for those physically unable to do things like grocery shop on their own, a robot could one day do it with them, or even for them. And as designers of autonomous vehicles, we are motivated by the opportunity to massively improve the safety of our roads the way we have in our skies.

Reaping the benefits of working robots is not a matter of launching the right start-up, however, or developing more sophisticated machine learning. Our relationship to technology will have to change. Collaborative robots will be fundamentally new societal entities, and they lack the millennia of cultural development that made us who we are as humans with incredibly advanced social skills. Making sure robots play well in the sandbox is a matter of safety, and it's a matter of sharing the benefits and mitigating the harms. Getting this right is so hard because it requires both technological and social revolutions.

The promise of human-robot partnerships is that we can offload tasks that, for a variety of reasons, robots are better at doing than humans. But our society is complex and impossible to capture fully in a computer model. Nor can we deterministically model the behaviors of robots in society well enough to predict and prevent all adverse behavior. Introducing robots introduces vulnerability and uncertainty into our lives, and so creating trusted partnerships must be our primary goal. We need to set higher standards for robot performance than for humans, because robots will be doing things that, at least sometimes, humans won't be capable of doing. When we remove people from the direct control of robots, we expose each and every one of us to new vulnerabilities. And the tech community must take this responsibility seriously. We've spent so much time in this book thinking about how advanced robots might break, and this is because human-robot partnerships will always be most successful when robots are designed to keep us cognizant of their limitations. But this is not at all how many workers in the tech community currently approach the design of robots. Rather, the obsessive focus has been on adding more features and making tech "fun." That approach is now a safety risk.

Instead of trying to develop perfect robots, we need to focus on developing perfect partnerships. To achieve this goal, we have argued, it is not feasible or even possible to design robots that are completely independent. We need layers of protection to act as checks and balances on these new social entities, the same way we help people all around us every day in both big and small ways.

Our aim in this book has been to provide some guidance on how to proceed in this moment in technology development, a moment that seems both familiar and completely alien. We all need to start figuring out the basic terms of the conversations we'll need to have, both in the near term and in the long term, about robots and society. And we need to figure out how to accommodate both the ramp-up of working robots in society *and* the learning curve these robots will be on.

The new era of working robots is familiar to us in that some branches of industrial design have long understood the value of designing perfect partnerships between automation and people, and their experiences offer a great deal of insight into building automated machines that can handle superhuman, safety-critical tasks. Decades of research into the fundamental strengths and limitations of humans to supervise automated systems provide us with a basis for understanding what is achievable, and what is not, as mobile command centers spin up to supervise the robots, drones, and self-driving cars that will blanket our cities. This prior research also gives us a preview of the pitfalls we can expect as we invite working robots into our homes to help with intimate tasks such as supervising our children. However, we quickly find ourselves in uncharted waters as these robots begin to interact and collaborate with bystanders and lay users, who will be essentially ignorant of how these machines work.

The way forward is to devote new effort to defining the roles of both supervisors and bystanders, to identify the dominant ways in which people will likely interact with these systems, and to design robots' user interfaces and intelligence to strengthen that partnership. That means, among many other things, making sure people are able to quickly understand robots' dominant modes. A supervisor, for example, needs an active mental model of a robot's functioning, and a means of predicting its future behavior, in order to be capable of intervening when needed. Ensuring that a supervisor builds and maintains the appropriate mental model may well mean choosing a design that requires user interaction even when it feels unnecessary, in tension with UX design principles for delighting the user. Designing working robots that know how to navigate around bystanders and, if needed, interact with them will require a new way of conceiving of the technology—but not only on the part of designers. The bystanders themselves will need to be able to quickly develop a passive mental model, with enough understanding to resolve interference or conflict with a working robot that is passing by and does not share their goals. Bystanders may have to take effective action when working robots interfere with their lives, and they won't necessarily have much time to figure out how. When a bystander has to intervene with a working robot, the automation affordances—both static and dynamic—will have to be obvious. The designers of these affordances need to make it perceptually evident how to influence the robot's behavior and actions. Affordances will have to support both instinctive and deliberative interactions.

All of this makes sense only if we start thinking of robots and people as fundamentally interdependent and design robots with this principle in mind from the ground up. Like the tip of an iceberg, the user interface alone only represents a small portion of the robot's capability for interdependence. We must go deeper to change the way we design robot intelligence and overall system architecture. If we make this leap, we can begin to build robots that actually know how to get help when they need it.

We can further improve our partnerships with robots by ensuring that robots help each other out. Left to their own devices, individual companies may not jump to invest in cross-platform coordination that supports the system of a competitor. Movements like V2X (vehicle-to-everything), which enables vehicles to communicate with the world around them, represent an early move in the right direction. Networked robot intelligence and data-collection backbones for working robots come with substantial concerns regarding privacy and security for consumers as well as companies. Nevertheless, a system that addresses these concerns and supports even limited information-sharing of the environment, as well as local negotiation and coordination of activities, can make working robots safer when deployed at scale.

Hard-learned lessons in aviation have proven that these social entities must be capable of communicating with each other as their presence in everyday life increases. This means communicating directly via micro-interactions to resolve momentary conflicts, negotiating among themselves to resolve more complex problems, and crowdsourcing to create a network of robot intelligence that enables them to formulate a more complete, robust understanding of the world. There are times when working robots will need to do things without human involvement, or at least with very little involvement, because they are better at certain tasks than humans will ever be. In these cases, the only thing that can really help them is another robot.

Industrial applications have also taught us how infrastructure can help working robots in big and small ways. Infrastructure can ease the technological challenges associated with getting robots to fit seamlessly into our world and make them more reliable and far safer. We need to organize, outfit, and adapt our world to reduce societal entropy for these new social entities, so they have a fair shot at accomplishing all that we will require of them.

Finally, industrial applications have taught us that we can accelerate our learning across the robotics industry if we freely share data about automation failures and errors when they occur, even when they don't result in an accident. With this kind of sharing, we can wake the sleeping giant and improve technology before we see catastrophic accidents. We propose expanding this approach to cover all aspects of the design and testing process, rather than waiting until after working robots are fielded. We can do this by creating a Community of Tomorrow test facility that allows designers to test their robots in complex, dynamic settings closely approximating real life. We can also establish an RSRS (Robot Safety Reporting System), so that people can anonymously report robot failures once they are roaming our world.

Working robots are coming. They could make life a lot easier for a lot of people. But nothing guarantees that sunny outlook. Our goal here is to begin the conversations that need to happen among citizens, policy makers, and industry. We believe the solution lies at this intersection of technology and society. We believe that it is possible to build working robots that are safe and productive if we pay attention to what we've learned before and figure out what lessons we still need to learn. We believe that what to expect when you're expecting robots is that they will not work *for* you anymore, but *with* you.

Acknowledgments

W E WOULD LIKE TO EXPRESS OUR DEEP GRATITUDE TO OUR editor, Eric Henney, for making this book possible in every way.

Thank you to our husbands, Bobby and Neel, for their love and partnership, and to our parents, April, Don, Dora, and George, for their love and inspiration.

Thank you to our mentors—especially those at MIT's Department of Aeronautics and Astronautics (AeroAstro) who trained us and did pioneering research in human-automation interaction: David Mindell, John Hansman, Laurence Young, Tom Sheridan, Jeffrey Hoffman, Dava Newman, Charles Oman, Andy Liu, Joseph Saleh, and Missy Cummings.

Thank you to all the researchers in the Interactive Robotics Group, and in particular those whose works make the building blocks of this book: Abhizna Butchibabu, Matthew Gombolay, Reymundo Gutierrez, Bradley Hayes, Been Kim, Joseph Kim, Kyle Kotowick, Przemyslaw

Lasota, Shen Li, Claudia Pérez D'Arpino, Ramya Ramakrishnan, Giancarlo Sturla, Mycal Tucker, Jessie Yang, and Chongjie Zhang.

Thank you to all the researchers and engineers in the Information and Cognition division at Draper Laboratory, and in particular Draper colleagues whose works contributed to the tenets conveyed in this book: Emily Vincent, Caroline Harriott, Jana Schwartz, Troy Lau, Megan Mitchell, and Brent Appleby.

Last, we would like to acknowledge the graduate students we have collaborated with over the past three years to refine the themes of this book: Nyoman Anjani, Rachel Cabosky, Phil Cotter, Joshua Creamer, Sarah Gonzalez, Bobby Holden, Shen Li, Sandro Salgueiro, Akash Shah, Mycal Tucker, Shane Vigil, Chris Fourie, Snehalkumar Gaikwad, Inderraj Grewal, Clement Li, Lindsay Sanneman, Kailah Snelgrove, Zachary Talus, Takaya Ukai, and Alison Yu.

Notes

INTRODUCTION

1. "IFR Forecast: 1.7 Million New Robots to Transform the World's Factories by 2020," International Federation of Robotics, press release, September 27, 2017, https://ifr.org/ifr-press-releases/news/ifr-forecast-1.7-million-new-robots-to -transform-the-worlds-factories-by-20.

2. "31 Million Robots Helping in Households Worldwide by 2019," International Federation of Robotics, press release, December 20, 2016, https:// ifr.org/ifr-press-releases/news/31-million-robots-helping-in-households -worldwide-by-2019.

3. "Road Safety Facts," Association for Safe International Road Travel, https:// www.asirt.org/safe-travel/road-safety-facts.

4. "Road Safety Facts," Association for Safe International Road Travel, https://www.asirt.org/safe-travel/road-safety-facts; Nathan Bomey, "2018 Was Third-Deadliest of the Decade on American Roads, NHTSA Says," *USA Today*, June 17, 2019, https://www.usatoday.com/story/money/cars/2019/06/17/car -crashes-36-750-people-were-killed-us-2018-nhtsa-estimates/1478103001.

5. "Aviation Safety Network Releases 2018 Airliner Accident Statistics," Flight Safety Foundation, press release, January 1, 2019, https://news.aviation-safety .net/2019/01/01/aviation-safety-network-releases-2018-airliner-accident-statistics.

CHAPTER 1. THE AUTOMATION INVASION

1. Janosch Delcker, "The Man Who Invented the Self-Driving Car (in 1986)," *Politico*, July 19, 2018, https://www.politico.eu/article/self-driving-car-born-1986 -ernst-dickmanns-mercedes.

2. "The DARPA Grand Challenge: Ten Years Later," Defense Advanced Research Projects Agency, press release, March 13, 2014, https://www.darpa.mil /news-events/2014-03-13.

3. Phil LeBeau, "Tesla Rolls Out Autopilot Technology," CNBC, October 14, 2015, https://www.cnbc.com/2015/10/14/tesla-rolls-out-autopilot-technology .html.

4. William Scheck, "Lawrence Sperry: Genius on Autopilot," *Aviation History* 15, no. 2 (2004): 46.

5. Calvin R. Jarvis, "Flight-Test Evaluation of an On-Off Rate Command Attitude Control System of a Manned Lunar-Landing Research Vehicle," *NASA Technical Note*, April 1967, NASA Technical Reports Server, https://ntrs.nasa.gov /archive/nasa/casi.ntrs.nasa.gov/19670013964.pdf.

6. Carl S. Droste and James E. Walker, *A Case Study on the F-16 Fly-by-Wire Flight Control System* (Reston, VA: American Institute of Aeronautics and Astronautics, 1985).

7. Sam Liden, "The Evolution of Flight Management Systems," in *AIAA/ IEEE Digital Avionics Systems Conference 13th DASC*, Phoenix, 1994, 157–169.

8. Alexis C. Madrigal, "Inside Google's Secret Drone-Delivery Program," *The Atlantic*, August 28, 2014, https://www.theatlantic.com/technology/archive/2014 /08/inside-googles-secret-drone-delivery-program/379306.

9. John Markoff, "Google's Next Phase in Driverless Cars: No Steering Wheel or Brake Pedals," *New York Times*, May 27, 2014, https://www.nytimes .com/2014/05/28/technology/googles-next-phase-in-driverless-cars-no-brakes-or -steering-wheel.html.

10. Charles S. Draper, H. P. Whitaker, and L. R. Young, "The Roles of Men and Instruments in Control and Guidance Systems for Spacecraft," in *15th International Astronautical Congress*, Poland, 1964.

11. Mark S. Young, Neville A. Stanton, and Don Harris, "Driving Automation: Learning from Aviation About Design Philosophies," *International Journal of Vehicle Design* 45, no. 3 (2007): 323–338.

12. Young et al., "Driving Automation."

13. David A. Mindell, *Digital Apollo: Human and Machine in Spaceflight* (Cambridge, MA: MIT Press, 2011).

14. Robert W. Bailey, "Performance vs. Preference," *Proceedings of the Human Factors and Ergonomics Society Annual Meeting* 37, no. 4 (October 1993): 282–286,

https://doi.org/10.1177/154193129303700406; Eric Frøkjær, Morten Hertzum, and Kasper Hornbæk, "Measuring Usability: Are Effectiveness, Efficiency, and Satisfaction Really Correlated?," in *Proceedings of the SIGCHI Conference on Human Factors in Computing Systems*, The Hague, 2000, 345–352, https://dl.acm.org/citation.cfm?id=332455.

CHAPTER 2. THERE IS NO SUCH THING AS A SELF-RELIANT ROBOT

1. S. M. K. Quadri and Sheikh Umar Farooq, "Software Testing—Goals, Principles, and Limitations," *International Journal of Computer Applications* 6, no. 9 (2010): 7–10, https://doi.org/10.5120/1343-1448.

2. John W. Senders and Neville P. Moray, *Human Error: Cause, Prediction, and Reduction* (Boca Raton, FL: CRC Press, 1995); James Reason, *Human Error* (Cambridge: Cambridge University Press, 1990).

3. James Reason, Erik Hollnagel, and Jean Paries, "Revisiting the Swiss Cheese Model of Accidents," *Journal of Clinical Engineering* 27, no. 4 (January 2006): 110–115.

4. Ralph T. Putnam, "Structuring and Adjusting Content for Students: A Study of Live and Simulated Tutoring of Addition," *American Educational Research Journal* 24, no. 1 (1987): 13–48, https://doi.org/10.3102/00028312024001013; Stephanie Ann Siler and Kurt VanLehn, "Investigating Microadaptation in One-to-One Human Tutoring," *Journal of Experimental Education* 83, no. 3 (July 2015): 344–367, https://doi.org/10.1080/00220973.2014.907224.

5. Derek Sleeman, Anthony E. Kelly, R. Martinak, R. D. Ward, and J. L. Moore, "Studies of Diagnosis and Remediation with High School Algebra Students," *Cognitive Science* 13, no. 4 (1989): 551–568, https://doi.org/10.1207/s15516709cog1304_3.

6. Julie A. Shah, Kevin A. Gluck, Tony Belpaeme, Kenneth R. Koedinger, Katharina J. Rohlfing, Han L. J. van der Maas, Paul Van Eecke, Kurt VanLehn, Anna-Lisa Vollmer, and Matthew Yee-King, "Task Instruction," in *Interactive Task Learning: Humans, Robots, and Agents Acquiring New Tasks Through Natural Interactions*, ed. Kevin A. Gluck and John E. Laird (Cambridge, MA: MIT Press, 2018), 169–192.

7. Robert L. Helmreich, "Building Safety on Three Cultures of Aviation," in *Proceedings of the IATA Human Factors Seminar*, Bangkok, 1998, 39–43, https://www.pacdeff.com/pdfs/3%20Cultures%20of%20Aviation%20Helmreich.pdf.

8. Ute Fischer, Judith Orasanu, and J. Kenneth Davison, "Cross-Cultural Barriers to Effective Communication in Aviation," in *Cross-Cultural Work Groups*, ed. Cherlyn S. Granrose and Stuart Oskamp (Thousand Oaks, CA: Sage, 1997).

9. Solace Shen, Hamish Tennent, Houston Claure, and Malte Jung, "My Telepresence, My Culture?," *Proceedings of the 2018 CHI Conference on Human*

Factors in Computing Systems, Montreal, 2018, https://doi.org/10.1145/3173574 .3173625.

10. Jennifer Shuttleworth, "SAE Standards News: J3016 Automated-Driving Graphic Update," SAE International, January 7, 2019, https://www.sae .org/news/2019/01/sae-updates-j3016-automated-driving-graphic.

11. Ramya Ramakrishnan, "Error Discovery Through Human-AI Collaboration" (PhD diss., Massachusetts Institute of Technology, 2019).

12. Janis Cannon-Bowers, Eduardo Salas, and Sharolyn Converse, "Shared Mental Models in Expert Team Decision Making," *Individual and Group Decision Making: Current Issues* 221 (1993): 221–246.

13. Rob Gray, Nancy Cooke, Nathan McNeese, and Jamie McNabb, "Investigating Team Coordination in Baseball Using a Novel Joint Decision Making Paradigm," *Frontiers in Psychology* 8, no. 907 (June 2017), https://doi.org/10.3389 /fpsyg.2017.00907.

14. Jamie C. Gorman, Nancy J. Cooke, and Polemnia G. Amazeen, "Training Adaptive Teams," *Human Factors* 52, no. 2 (July 2010), https://doi.org /10.1177/0018720810371689.

15. Ramya Ramakrishnan, Chongjie Zhang, and Julie Shah, "Perturbation Training for Human-Robot Teams," *Journal of Artificial Intelligence Research* 59 (July 2017): 495–541, https://doi.org/10.1613/jair.5390.

16. Jeff Zacks and Barbara Tversky, "Bars and Lines: A Study of Graphic Communication," *Memory and Cognition* 27, no. 6 (1999): 1073–1079, https:// doi.org/10.3758/bf03201236.

17. Lance Sherry, Karl Fennell, Michael Feary, and Peter Polson, "Human-Computer Interaction Analysis of Flight Management System Messages," *Journal of Aircraft* 43, no. 5 (2006): 1372–1376, https://doi.org/10.2514/1.20026.

18. Kim J. Vicente and Jens Rasmussen, "Ecological Interface Design: Theoretical Foundations," *IEEE Transactions on Systems, Man, and Cybernetics* 22, no. 4 (July 1992): 589–606, https://doi.org/10.1109/21.156574.

19. William S. Pawlak and Kim J. Vicente, "Inducing Effective Operator Control Through Ecological Interface Design," *International Journal of Human-Computer Studies* 44, no. 5 (1996): 653–688; Michael E. Janzen and Kim J. Vicente, "Attention Allocation Within the Abstraction Hierarchy," *International Journal of Human-Computer Studies* 48, no. 4 (1998): 521–545.

20. Natalie Wolchover, "Breaking the Code: Why Yuor Barin Can Raed Tihs," *Live Science*, February 9, 2012, https://www.livescience.com/18392-reading -jumbled-words.html.

21. Caprice C. Greenberg, Scott E. Regenbogen, Stuart R. Lipsitz, Rafael Diaz-Flores, and Atul A. Gawande, "The Frequency and Significance of Discrepancies in the Surgical Count," *Annals of Surgery* 248, no. 2 (2008): 337–341.

22. Mark A. Staal, "Stress, Cognition, and Human Performance: A Literature Review and Conceptual Framework," *NASA Technical Memorandum*, August 2004, available from NASA Technical Reports Server, https://https://ntrs .nasa.gov/archive/nasa/casi.ntrs.nasa.gov/20060017835.pdf.

23. David E. Kieras and Susan Bovair, "The Role of a Mental Model in Learning to Operate a Device," *Cognitive Science* 8, no. 3 (July 1984), https://doi .org/10.1016/S0364-0213(84)80003-8.

CHAPTER 3. WHEN ROBOTS ARE TOO GOOD

1. *Final Report on the Accident on 1st June 2009 to the Airbus A330-203 Registered F-GZCP Operated by Air France Flight AF 447 Rio de Janeiro–Paris* (Le Bourget, France: Bureau d'Enquêtes et d'Analyses pour la sécurité de l'aviation civile, July 2012), https://www.bea.aero/docspa/2009/f-cp090601.en/pdf /f-cp090601.en.pdf.

2. *Aircraft Accident Report: Eastern Airlines 401/l-1011, Miami, FL, 29 December 1972*, NTSB/AAR-73-14 (Washington, DC: National Transportation Safety Board, 1973); *Aircraft Accident Report: United Airlines, Inc., McDonnell Douglas DC-8-61, N8082U, Portland, OR, 28 December 1978*, NTSB/AAR-79-07 (Washington, DC: National Transportation Safety Board, 1979); *Aircraft Separation Incidents at Hartsfield Atlanta International Airport, Atlanta, GA*, NTSB/SIR-81-6 (Washington, DC: National Transportation Safety Board, 1981); *Aircraft Accident Report: Northwest Airlines, Inc., McDonnell-Douglas DC-9-82, N312RC, Detroit Metropolitan Wayne County Airport, 16 August 1987*, NTSB/AAR-99-05 (Washington, DC: National Transportation Safety Board, 1988); *Aircraft Accident Report: US Air Flight 105, Boeing 737-200, N282AU, Kansas International Airport, MO, 8 September 1989*, NTSB/AAR-90-04 (Washington, DC: National Transportation Safety Board, 1990).

3. Mica R. Endsley and David B. Kaber, "Level of Automation Effects on Performance, Situation Awareness and Workload in a Dynamic Control Task," *Ergonomics* 42, no. 3 (March 1999): 462–492.

4. *Traffic Safety Facts: Crash Stats* (Washington, DC: US Department of Transportation, National Highway Traffic Safety Administration, February 2015), https://crashstats.nhtsa.dot.gov/Api/Public/ViewPublication/812115.

5. *Final Report on the Accident on 1st June 2009 to the Airbus A330-203 Registered F-GZCP Operated by Air France Flight AF 447 Rio de Janeiro–Paris*.

6. *Managing Human Error*, postnote No. 156 (London: Parliamentary Office of Science and Technology, June 2001), https://www.parliament.uk/documents /post/pn156.pdf.

7. Joseph Stromberg, "Is GPS Ruining Our Ability to Navigate for Ourselves?," *Vox*, September 2, 2015, https://www.vox.com/2015/9/2/9242049/gps-maps

-navigation; Toru Ishikawa, Hiromichi Fujiwara, Osamu Imai, and Atsuyuki Okabe, "Wayfinding with a GPS-Based Mobile Navigation System: A Comparison with Maps and Direct Experience," *Journal of Environmental Psychology* 28, no. 1 (2008): 74–82.

8. Raja Parasuraman and Victor Riley, "Humans and Automation: Use, Misuse, Disuse, Abuse," *Human Factors* 39, no. 2 (June 1997): 230–253, https:// doi.org/10.1518/001872097778543886; Victor Riley, "Operator Reliance on Automation: Theory and Data," *Automation and Human Performance: Theory and Applications* (1996): 19–35.

9. V. Riley, "A Theory of Operator Reliance on Automation," in *Human Performance in Automated Systems: Recent Research and Trends*, ed. M. Mouloua and R. Parasuraman (Hillsdale, NJ: Erlbaum, 1994), 8–14.

10. Rebecca A. Grier, Joel S. Warm, William N. Dember, Gerald Matthews, Traci L. Galinsky, and Raja Parasuraman, "The Vigilance Decrement Reflects Limitations in Effortful Attention, Not Mindlessness," *Human Factors* 45, no. 3 (Fall 2013): 349–359, https://doi.org/10.1518/hfes.45.3.349.27253.

11. Paul Robinette, Wenchen Li, Robert Allen, Ayanna M. Howard, and Alan R. Wagner, "Overtrust of Robots in Emergency Evacuation Scenarios," *11th ACM/ IEEE International Conference on Human-Robot Interaction (HRI)*, Christchurch, 2016, https://doi.org/10.1109/hri.2016.7451740.

12. Deborah Baer, "What E.R. Doctors Wish You Knew," *Parents*, June 1, 2003, https://www.parents.com/health/doctors/what-er-doctors-wish-you-knew.

13. Barbara A. Morrongiello, "Caregiver Supervision and Child-Injury Risk: I. Issues in Defining and Measuring Supervision; II. Findings and Directions for Future Research," *Journal of Pediatric Psychology* 30, no. 7 (2005): 536–552, https:// doi.org/10.1093/jpepsy/jsi041.

14. Mica R. Endsley, "A Taxonomy of Situation Awareness Errors," *Human Factors in Aviation Operations* 3, no. 2 (1995): 287–292.

15. Amos Tversky and Daniel Kahneman, "Judgment Under Uncertainty: Heuristics and Biases," *Science* 185, no. 4157 (1974): 1124–1131, https://doi.org /10.1126/science.185.4157.1124.

16. Herbert A. Simon, "A Behavioral Model of Rational Choice," *Quarterly Journal of Economics* 69, no. 1 (1955): 99, https://doi.org/10.2307/1884852; Anthony D. Cox and John O. Summers, "Heuristics and Biases in the Intuitive Projection of Retail Sales," *Journal of Marketing Research* 24, no. 3 (1987): 290, https://doi.org/10.2307/3151639; Brian Smith, Priya Sharma, and Paula Hooper, "Decision Making in Online Fantasy Sports Communities," *Interactive Technology and Smart Education* 3, no. 4 (2006): 347–360, https://doi.org /10.1108/17415650680000072; Gary Klein, Roberta Calderwood, and Anne Clinton-Cirocco, "Rapid Decision Making on the Fire Ground: The Original

Study Plus a Postscript," *Journal of Cognitive Engineering and Decision Making* 4, no. 3 (2010): 186–209, https://doi.org/10.1518/155534310x12844000801203; Judith M. Orasanu, "Flight Crew Decision-Making," in *Crew Resource Management*, 2nd ed., ed. Barbara G. Kanki, Robert L. Helmreich, and Jose Anca (San Diego: Elsevier, 2010), 147–179, https://pdfs.semanticscholar.org/442d /0b8e4c936b84f15d2814dd0871fdc896f40f.pdf.

17. Morrongiello, "Caregiver Supervision and Child-Injury Risk."

18. Morrongiello, "Caregiver Supervision and Child-Injury Risk."

19. Alice LaPlante, "Robot Nannies Are Here, but Won't Replace Your Babysitter—Yet," *Forbes*, March 29, 2017, https://www.forbes.com/sites/century link/2017/03/29/robot-nannies-are-here-but-wont-replace-your-babysitter-yet /#37511fb956b7.

20. Mica R. Endsley, "Toward a Theory of Situation Awareness in Dynamic Systems," *Human Factors* 37, no. 1 (March 1995): 32–64, https://doi.org/10.1518 /001872095779049543.

21. *Aircraft Accident Report, China Airlines Boeing 747-SP, N4522V 300 Nautical Miles Northwest of San Francisco, CA, February 19, 1985, NTSB/AAR-86/03* (Washington, DC: National Transportation Safety Board, 1986), https:// www.ntsb.gov/investigations/AccidentReports/Reports/AAR8603.pdf.

22. Robert M. Yerkes and John D. Dodson, "The Relation of Strength of Stimulus to Rapidity of Habit-Formation," *Journal of Comparative Neurology and Psychology* 18, no. 5 (1908): 459–482, https://doi.org/10.1002/cne.920180503.

23. A. O'Dhaniel, Ruth L. F. Leong, and Yoanna A. Kurnianingsih, "Cognitive Fatigue Destabilizes Economic Decision Making Preferences and Strategies," *PLOS One* 10, no. 7 (2015); Davide Dragone, "I Am Getting Tired: Effort and Fatigue in Intertemporal Decision-Making," *Journal of Economic Psychology* 30, no. 4 (2009): 552–562; Linda D. Scott, Cynthia Arslanian-Engoren, and Milo C. Engoren, "Association of Sleep and Fatigue with Decision Regret Among Critical Care Nurses," *American Journal of Critical Care* 23, no. 1 (2014): 13–23; Mitchell R. Smith, Linus Zeuwts, Matthieu Lenoir, Nathalie Hens, Laura M. S. De Jong, and Aaron J. Coutts, "Mental Fatigue Impairs Soccer-Specific Decision-Making Skill," *Journal of Sports Sciences* 34, no. 14 (2016): 1297–1304; Shai Danziger, Jonathan Levav, and Liora Avnaim-Pesso, "Extraneous Factors in Judicial Decisions," *Proceedings of the National Academy of Sciences* 108, no. 17 (2011): 6889–6892.

24. Joshua Rubinstein, David E. Meyer, and Jeffrey E. Evans, "Executive Control of Cognitive Processes in Task Switching," *Journal of Experimental Psychology: Human Perception and Performance* 27, no. 4 (August 2001): 763–797, https://doi.org/10.1037/0096-1523.27.4.763.

25. Jerry L. Franke, Jody J. Daniels, and Daniel C. Mcfarlane, "Recovering Context After Interruption," *Proceedings of the Twenty-Fourth Annual Conference*

of the Cognitive Science Society (New York: Routledge, 2019), 310–315, https://doi.org/10.4324/9781315782379-90; Kyle Kotowick and Julie Shah, "Intelligent Sensory Modality Selection for Electronic Supportive Devices," in *Proceedings of the 22nd International Conference on Intelligent User Interfaces*, 2017, 55–66.

26. R. John Hansman, "Complexity in Aircraft Automation—A Precursor for Concerns in Human-Automation Systems," *Phi Kappa Phi Journal* 81, no. 1 (2001): 30; Michael A. Mollenhauer, Thomas A. Dingus, Cher Carney, Jonathan M. Hankey, and Steve Jahns, "Anti-Lock Brake Systems: An Assessment of Training on Driver Effectiveness," *Accident Analysis and Prevention* 29, no. 1 (1997): 97–108, https://doi.org/10.1016/s0001-4575(96)00065-6.

27. Nadine B. Sarter and David D. Woods, "How in the World Did We Ever Get into That Mode? Mode Error and Awareness in Supervisory Control," *Human Factors* 37, no. 1 (1995): 5–19, https://doi.org/10.1518/001872095779049516.

28. Mica R. Endsley, "Autonomous Driving Systems: A Preliminary Naturalistic Study of the Tesla Model S," *Journal of Cognitive Engineering and Decision Making* 11, no. 3 (2017): 225–238, https://doi.org/10.1177/1555343417695197.

29. *Collision Between US Navy Destroyer John S McCain and Tanker Alnic MC Singapore Strait, 5 Miles Northeast of Horsburgh Lighthouse, August 21, 2017, NTSB/MAR-19/01 PB2019-100970* (Washington, DC: National Transportation Safety Board, 2019).

30. Sanjay S. Vakil and R. John Hansman Jr., "Approaches to Mitigating Complexity-Driven Issues in Commercial Autoflight Systems," *Reliability Engineering and System Safety* 75, no. 2 (2002): 133–145.

31. Daniel Kahneman and Amos Tversky, "Choices, Values, and Frames," *American Psychologist* 39, no. 4 (1984): 341–350, https://doi.org/10.1037/0003-066X.39.4.341.

32. J. R. Hackman, *Leading Teams: Setting the Stage for Great Performances* (Boston: Harvard Business School Press, 2002); George E. Cooper, Maurice D. White, and John K. Lauber, "Resource Management on the Flightdeck," NASA Conference Publication 2120 (Moffett Field, CA: NASA, 1980), available from NASA Technical Reports Server, https://ntrs.nasa.gov/archive/nasa/casi.ntrs.nasa.gov/19800013796.pdf.

33. Matthew C. Gombolay, Reymundo A. Gutierrez, Shanelle G. Clarke, Giancarlo F. Sturla, and Julie A. Shah, "Decision-Making Authority, Team Efficiency and Human Worker Satisfaction in Mixed Human–Robot Teams," *Autonomous Robots* 39, no. 3 (2015): 293–312, https://doi.org/10.1007/s10514-015-9457-9.

34. Gombolay et al., "Decision-Making Authority."

35. Jimmy Baraglia, Maya Cakmak, Yukie Nagai, Rajesh Rao, and Minoru Asada, "Initiative in Robot Assistance During Collaborative Task Execution," in *11th ACM/IEEE International Conference on Human Robot Interaction*, Christ-

church, 2016, 67–74; Guy Hoffman and Cynthia Breazeal, "Effects of Antici-patory Action on Human-Robot Teamwork Efficiency, Fluency, and Perception of Team," in *2nd ACM/IEEE International Conference on Human-Robot Inter-action*, Arlington, VA, 2007, 1–8, https://doi.org/10.1145/1228716.1228718; Chien-Ming Huang and Bilge Mutlu, "Anticipatory Robot Control for Efficient Human-Robot Collaboration," in *11th ACM/IEEE International Conference on Human Robot Interaction*, Christchurch, 2016, 83–90; Chang Liu, Jessica Ham-rick, Jaime Fisac, Anca Dragan, J. Hedrick, S. Sastry, and Thomas Griffiths, "Goal Inference Improves Objective and Perceived Performance in Human-Robot Col-laboration," in *AAMAS 2016: International Conference on Autonomous Agents and Multiagent Systems*, Singapore, 2016, 940–948.

36. Jessie Y. Chen, Katelyn Procci, Michael Boyce, Julia Wright, Andre Gar-cia, and Michael Barnes, *Situation Awareness–Based Agent Transparency*, No. ARL-TR-6905, Army Research Lab, Aberdeen Proving Ground, Maryland, Hu-man Research and Engineering Directorate, 2014.

37. Endsley, "Autonomous Driving Systems."

38. David B. Kaber and Mica R. Endsley, "The Effects of Level of Automa-tion and Adaptive Automation on Human Performance, Situation Awareness and Workload in a Dynamic Control Task," *Theoretical Issues in Ergonomics Science* 5, no. 2 (2004): 153.

CHAPTER 4. THE THREE-BODY PROBLEM

1. Matt Simon, "San Francisco Just Put the Brakes on Delivery Robots," *Wired Science*, December 6, 2017, https://www.wired.com/story/san-francisco -just-put-the-brakes-on-delivery-robots.

2. Donald Norman, *The Design of Everyday Things* (New York: Doubleday, 1988).

3. Josh Hrala, "A Mall Security Robot Has Knocked Down and Run Over a Toddler in Silicon Valley," *Science Alert*, July 15, 2016, https://www.sciencealert .com/a-mall-security-robot-recently-knocked-down-and-ran-over-a-toddler-in -silicon-valley.

4. Walter J. Freeman, "Comparison of Brain Models for Active vs. Passive Perception," *Information Sciences* 116, nos. 2–4 (1999): 97–107, https://doi.org /10.1016/s0020-0255(98)10100-7.

5. Dorsa Sadigh, Shankar Sastry, Sanjit A. Seshia, and Anca D. Dragan, "Plan-ning for Autonomous Cars That Leverage Effects on Human Actions," *Robotics: Science and Systems* 12 (2016), https://doi.org/10.15607/rss.2016.xii.029; Sonia Chernova, Vivian Chu, Angel Daruna, Haley Garrison, Meera Hahn, Priyanka Khante, Weiyu Liu, and Andrea Thomaz, "Situated Bayesian Reasoning Framework

for Robots Operating in Diverse Everyday Environments," in *Robotics Research: The 18th International Symposium ISRR*, ed. Nancy M. Amato, Greg Hager, Shawna Thomas, and Miguel Torres-Torriti, Springer Proceedings in Advanced Robotics, vol. 10 (Cham, Switzerland: Springer, 2020), 353–369, https://doi.org/10.1007/978-3 -030-28619-4_29; Micah Carroll, Rohin Shah, Mark K. Ho, Tom Griffiths, Sanjit Seshia, Pieter Abbeel, and Anca Dragan, "On the Utility of Learning About Humans for Human-AI Coordination," in *Advances in Neural Information Processing Systems* (2019): 5175–5186; Rohan Paul, Jacob Arkin, Derya Aksaray, Nicholas Roy, and Thomas M. Howard, "Efficient Grounding of Abstract Spatial Concepts for Natural Language Interaction with Robot Platforms," *International Journal of Robotics Research* 37, no. 10 (2018): 1269–1299; David Whitney, Miles Eldon, John Oberlin, and Stefanie Tellex, "Interpreting Multimodal Referring Expressions in Real Time," in *2016 IEEE International Conference on Robotics and Automation (ICRA)*, Stockholm, 2016, 3331–3338, https://doi.org/10.1109/icra.2016.7487507; Karol Hausman, Yevgen Chebotar, Stefan Schaal, Gaurav Sukhatme, and Joseph J. Lim, "Multi-Modal Imitation Learning from Unstructured Demonstrations Using Generative Adversarial Nets," in *NIPS'17: Advances in Neural Information Processing Systems* (December 2017), 1235–1245; Matthew Gombolay, Reed Jensen, Jessica Stigile, Toni Golen, Neel Shah, Sung-Hyun Son, and Julie Shah, "Human-Machine Collaborative Optimization via Apprenticeship Scheduling," *Journal of Artificial Intelligence Research* 63 (2018): 1–49, https://doi.org/10.1613/jair.1.11233; Emmanuel Senft, Séverin Lemaignan, Paul E. Baxter, Madeleine Bartlett, and Tony Belpaeme, "Teaching Robots Social Autonomy from in Situ Human Guidance," *Science Robotics* 4, no. 35 (2019), https://doi.org/10.1126/scirobotics.aat1186; Manuela M. Veloso, Joydeep Biswas, Brian Coltin, and Stephanie Rosenthal, "CoBots: Robust Symbiotic Autonomous Mobile Service Robots," in *IJCAI'15: Proceedings of the 24th International Conference on Artificial Intelligence* (Palo Alto, CA: AAAI Press, 2015), 4423.

6. "Hospital Recruits Robot Cleaner," BBC News, June 19, 2001, http://news .bbc.co.uk/2/hi/health/1396433.stm.

7. Bilge Mutlu and Jodi Forlizzi, "Robots in Organizations: The Role of Workflow, Social, and Environmental Factors in Human-Robot Interaction," in *3rd ACM/IEEE International Conference on Human Robot Interaction*, Amsterdam, 2008, 287–294.

8. Michael Leonard, Suzanne Graham, and Doug Bonacum, "The Human Factor: The Critical Importance of Effective Teamwork and Communication in Providing Safe Care," *BMJ Quality and Safety* 13, suppl. 1 (2004): i85–i90; "Sentinel Event," Joint Commission, https://www.jointcommission.org/resources/patient -safety-topics/sentinel-event; "Patient Safety," Joint Commission, https://www .jointcommission.org/resources/patient-safety-topics/patient-safety.

9. M. Gombolay, X. Jessie Yang, B. Hayes, N. Seo, Z. Liu, S. Wadhwania, T. Yu, N. Shah, T. Golen, and J. Shah, "Robotic Assistance in Coordination of Patient Care," in *Robotics: Science and Systems* 12 (June 2016), https://doi.org/10 .15607/RSS.2016.XII.026.

10. Wendy Ju, "The Design of Implicit Interactions," *Synthesis Lectures on Human-Centered Informatics* 8, no. 2 (2015): 1–93, https://doi.org/10.2200 /s00619ed1v01y201412hci028.

11. Nellie Bowles, "Google Self-Driving Car Collides with Bus in California, Accident Report Says," *The Guardian*, March 1, 2016, https://www.theguardian .com/technology/2016/feb/29/google-self-driving-car-accident-california.

12. Cynthia Breazeal, "Social Interactions in HRI: The Robot View," *IEEE Transactions on Systems, Man, and Cybernetics, Part C: Applications and Reviews* 34, no. 2 (2004): 181–186; Hee Rin Lee and Selma Sabanović, "Culturally Variable Preferences for Robot Design and Use in South Korea, Turkey, and the United States," in *9th ACM/IEEE International Conference on Human-Robot Interaction (HRI)*, Bielefeld, Germany, 2014, 17–24, https://doi.org/10.1145 /2559636.2559676.

13. John D. Lee and Katrina A. See, "Trust in Automation: Designing for Appropriate Reliance," *Human Factors* 46, no. 1 (2004): 50–80, https://doi .org/10.1518/hfes.46.1.50.30392.

14. Kristin E. Schaefer, Jessie Y. C. Chen, James L. Szalma, and P. A. Hancock, "A Meta-Analysis of Factors Influencing the Development of Trust in Automation," *Human Factors* 58, no. 3 (2016): 377–400, https://doi.org/10.1177 /0018720816634228; Peter A. Hancock, Deborah R. Billings, Kristin E. Schaefer, Jessie Y. C. Chen, Ewart J. De Visser, and Raja Parasuraman, "A Meta-Analysis of Factors Affecting Trust in Human-Robot Interaction," *Human Factors* 53, no. 5 (2011): 517–527, https://doi.org/10.1177/0018720811417254; Jessie X. Yang, V. V. Unhelkar, K. Li, and J. A. Shah, "Evaluating Effects of User Experience and System Transparency on Trust in Automation," in *12th ACM/IEEE International Conference on Human Robot Interaction (HRI)*, Vienna, 2017; Shih-Yi Chien, Michael Lewis, Katia Sycara, Asiye Kumru, and Jyi-Shane Liu, "Influence of Culture, Transparency, Trust, and Degree of Automation on Automation Use," *IEEE Transactions on Human-Machine Systems* (2019).

15. Yang et al., "Evaluating Effects of User Experience."

16. M. C. Gombolay, R. A. Gutierrez, S. G. Clarke, G. F. Sturla, and J. A. Shah, "Decision-Making Authority, Team Efficiency and Human Worker Satisfaction in Mixed Human-Robot Teams," *Autonomous Robots* 39, no. 3 (October 2015): 312.

17. Norman, *Design of Everyday Things*.

18. Bob Johnson, "Man Killed in Crash; Drone Operator May Face Charges for Flying over Scene," *mlive*, August 18, 2017, https://www.mlive.com/news /saginaw/2017/08/man_killed_in_crash_drone_oper.html.

19. "Drone Citings and Near Misses," Center for the Study of the Drone at Bard College, August 2015, https://dronecenter.bard.edu/files/2015/08/Near -Misses-Clean-Version-6.pdf.

CHAPTER 5. ROBOTS DON'T HAVE TO BE CUTE

1. Candace Lombardi, "Are Drivers Ready for High-Tech Onslaught?," *Roadshow by CNET*, August 29, 2007, https://www.cnet.com/roadshow/news/are -drivers-ready-for-high-tech-onslaught.

2. Donald A. Norman, "Interaction Design for Automobile Interiors," November 17, 2008, https://jnd.org/interaction_design_for_automobile_interiors.

3. Alessandro Giusti, Jérôme Guzzi, Dan C. Cireşan, Fang-Lin He, Juan P. Rodríguez, Flavio Fontana, Matthias Faessler, et al., "A Machine Learning Approach to Visual Perception of Forest Trails for Mobile Robots," *IEEE Robotics and Automation Letters* 1, no. 2 (July 2016): 661–667, https://doi.org/10.1109 /LRA.2015.2509024.

4. Diane Coutu, "Why Teams Don't Work," *Harvard Business Review*, May 2009, https://hbr.org/2009/05/why-teams-dont-work.

5. Jakob Nielsen and Rolf Molich, "Heuristic Evaluation of User Interfaces," *Proceedings of the SIGCHI Conference on Human Factors in Computing Systems*, Seattle, 1990, 249–256, https://doi.org/10.1145/97243.97281.

6. "MHCI Curriculum," Carnegie Mellon University Human-Computer Interaction Institute, https://www.hcii.cmu.edu/academics/mhci/core-curriculum.

7. Jake Knapp, with John Zeratsky and Braden Kowitz, *Sprint: How to Solve Big Problems and Test New Ideas in Just Five Days* (New York: Simon and Schuster, 2016).

8. "Make Your UX Design Process Agile Using Google's Methodology," Interaction Design Foundation, https://www.interaction-design.org/literature/article/make -your-ux-design-process-agile-using-google-s-methodology.

9. Bjorn Fehrm, "Aircraft Programme Cost," *Leeham News and Analysis*, March 21, 2016, https://leehamnews.com/2016/03/21/aircraft-programme-cost.

10. "iRobot 510 PackBot Multi-Mission Robot," *Army Technology*, https:// www.army-technology.com/projects/irobot-510-packbot-multi-mission-robot.

11. "The MIT DARPA Robotics Challenge Team," http://drc.mit.edu.

12. Matthew Johnson, Jeffrey M. Bradshaw, Paul J. Feltovich, Robert R. Hoffman, Catholijn Jonker, Birna van Riemsdijk, and Maarten Sierhuis, "Beyond

Cooperative Robotics: The Central Role of Interdependence in Coactive Design," *IEEE Intelligent Systems* 26, no. 3 (2011): 81–88; Martijn Ijtsma, Lanssie M. Ma, Amy R. Pritchett, and Karen M. Feigh, "Computational Methodology for the Allocation of Work and Interaction in Human-Robot Teams," *Journal of Cognitive Engineering and Decision Making* 13, no. 4 (2019): 221–241, https://doi.org/10.1177/1555343419869484.

13. Matthew Johnson, Jeffrey M. Bradshaw, Paul J. Feltovich, Catholijn M. Jonker, M. Birna Van Riemsdijk, and Maarten Sierhuis, "Coactive Design: Designing Support for Interdependence in Joint Activity," *Journal of Human-Robot Interaction* 3, no. 1 (2014): 43–69.

14. Jessie Y. C. Chen, Katelyn Procci, Michael Boyce, Julia Wright, Andre Garcia, and Michael Barnes, *Situation Awareness–Based Agent Transparency*, No. ARL-TR-6905 (Aberdeen, MD: Army Research Lab Human Research and Engineering Directorate, 2014); Jessie Y. C. Chen, Michael J. Barnes, Julia L. Wright, Kimberly Stowers, and Shan G. Lakhmani, "Situation Awareness–Based Agent Transparency for Human-Autonomy Teaming Effectiveness," *Micro- and Nanotechnology Sensors, Systems, and Applications IX* 101941V (2017), https://doi.org/10.1117/12.2263194.

15. Chen et al., *Situation Awareness–Based Agent Transparency*; Chen et al., "Situation Awareness–Based Agent Transparency for Human-Autonomy Teaming Effectiveness."

16. Claudia Pérez D'Arpino, "Hybrid Learning for Multi-Step Manipulation in Collaborative Robotics" (PhD diss., Massachusetts Institute of Technology, 2019), https://dspace.mit.edu/handle/1721.1/122740.

17. Maria Cvach, "Monitor Alarm Fatigue: An Integrative Review," *Biomedical Instrumentation and Technology* 46, no. 4 (2012): 268–277.

18. Michael J. Muller, "Participatory Design: The Third Space in HCI," in *The Human-Computer Interaction Handbook*, ed. Julie A. Jacko (Boca Raton, FL: CRC Press, 2007), 1087–1108.

19. Robert C. Martin, *Agile Software Development: Principles, Patterns, and Practices* (Harlow, UK: Pearson, 2002); Frauke Paetsch, Armin Eberlein, and Frank Maurer, "Requirements Engineering and Agile Software Development," in *WET ICE 2003, Proceedings, Twelfth IEEE International Workshops on Enabling Technologies: Infrastructure for Collaborative Enterprises* (Piscataway, NJ: IEEE, 2003), 308–313.

20. Hugh Beyer and Karen Holtzblatt, *Contextual Design: Defining Customer-Centered Systems* (London: Academic Press, 1998).

21. Nielsen and Molich, "Heuristic Evaluation of User Interfaces"; Clayton Lewis and Cathleen Wharton, "Cognitive Walkthroughs," in *Handbook of*

Human-Computer Interaction, ed. Marting G. Helander, Thomas K. Landauer, and Prasad V. Prabhu (Amsterdam: Elsevier B.V., 1997), https://doi.org/10.1016/B978-044481862-1.50096-0.

22. David E. Kieras, "Towards a Practical GOMS Model Methodology for User Interface Design," in *Handbook of Human-Computer Interaction*, ed. Marting G. Helander (Amsterdam: Elsevier B.V., 1988), https://doi.org/10.1016/B978-0-444-70536-5.50012-9.

23. John R. Anderson, "ACT: A Simple Theory of Complex Cognition," *American Psychologist* 51, no. 4 (1996): 355–365, https://doi.org/10.1037/0003-066x.51.4.355; Jerome R. Busemeyer and Adele Diederich, *Cognitive Modeling* (Thousand Oaks, CA: Sage, 2010).

24. Janni Nielsen, Torkil Clemmensen, and Carsten Yssing, "Getting Access to What Goes on in People's Heads?," *Proceedings of the Second Nordic Conference on Human-Computer Interaction (NordiCHI 02)* (New York: ACM, 2002), https://doi.org/10.1145/572020.572033; Shawn Patton, "The Definitive Guide to Playtest Questions," *Schell Games*, April 27, 2017, https://www.schellgames.com/blog/the-definitive-guide-to-playtest-questions.

25. Jen Cardello, "Three Uses for Analytics in User-Experience Practice," Nielsen Norman Group, November 17, 2013, https://www.nngroup.com/articles/analytics-user-experience/?lm=analytics-and-user-experience&pt=course.

CHAPTER 6. HOW DO YOU SAY "EXCUSE ME" TO A ROBOT?

1. Judea Pearl, "The Seven Tools of Causal Inference, with Reflections on Machine Learning," *Communications of the ACM* 62, no. 3 (2019): 54–60, https://doi.org/10.1145/3241036.

2. Finale Doshi-Velez and Been Kim, "Towards a Rigorous Science of Interpretable Machine Learning," *arXiv:1702.08608* (2017).

3. Daniel Kahneman, *Thinking, Fast and Slow* (New York: Farrar, Straus & Giroux, 2011).

4. Mycal Tucker and J. Shah, "Adversarially Guided Self-Play for Adopting Social Conventions," *arXiv:2001.05994* (2020); Bradley Hayes and Julie A. Shah, "Interpretable Models for Fast Activity Recognition and Anomaly Explanation During Collaborative Robotics Tasks," in *2017 IEEE International Conference on Robotics and Automation (ICRA)*, Marina Bay Sands, Singapore, 2017, https://doi.org/10.1109/icra.2017.7989778; Jyoti Aneja, Harsh Agrawal, Dhruv Batra, and Alexander Schwing, "Sequential Latent Spaces for Modeling the Intention During Diverse Image Captioning," *Proceedings of the IEEE International Conference on Computer Vision* (Piscataway, NJ: IEEE, 2019), 4261–4270.

5. Vaibhav Unhelkar, S. Li, and J. Shah, "Decision-Making for Bidirectional Communication in Sequential Human-Robot Collaborative Tasks," in *15th ACM/ IEEE International Conference on Human-Robot Interaction (HRI)*, Cambridge, 2020.

6. Terrence Fong, Illah Nourbakhsh, and Kerstin Dautenhahn, "A Survey of Socially Interactive Robots," *Robotics and Autonomous Systems* 42, nos. 3–4 (2003): 143–166; Jakub Złotowski, Diane Proudfoot, Kumar Yogeeswaran, and Christoph Bartneck, "Anthropomorphism: Opportunities and Challenges in Human–Robot Interaction," *International Journal of Social Robotics* 7, no. 3 (2015): 347–360.

7. Byron Reeves and Clifford Ivar Nass, *The Media Equation: How People Treat Computers, Television, and New Media Like Real People and Places* (Cambridge: Cambridge University Press, 1996).

8. Gary Klein, "The Recognition-Primed Decision (RPD) Model: Looking Back, Looking Forward," in *Naturalistic Decision Making*, ed. C. E. Zsambok and G. Klein (Hillsdale, NJ: Erlbaum, 1997), 285–292.

9. M. A. Brewer, K. Fitzpatrick, J. A. Whitacre, and D. Lord, "Exploration of Pedestrian Gap-Acceptance Behavior at Selected Locations," *Transportation Research Record* 1982, no. 1 (2006): 132–140.

10. Fong et al., "Survey of Socially Interactive Robots."

11. Przemyslaw A. Lasota and Julie A. Shah, "Analyzing the Effects of Human-Aware Motion Planning on Close-Proximity Human–Robot Collaboration," *Human Factors* 57, no. 1 (2015): 21–33, https://doi.org/10.1177/0018720814565188.

12. Herbert A. Simon and Alan Newell, *Human Problem Solving* (Englewood Cliffs, NJ: Prentice-Hall, 1972); J. Chang, J. Boyd-Graber, S. Gerrish, C. Wang, and D. M. Blei, "Reading Tea Leaves: How Humans Interpret Topic Models," in *NIPS'09: Proceedings of the 22nd International Conference on Neural Information Processing Systems* (2009), 288–296; Gary A. Klein, *A Recognition-Primed Decision (RPD) Model of Rapid Decision Making* (New York: Ablex, 1993).

13. Been Kim, Cynthia Rudin, and Julie Shah, "The Bayesian Case Model: A Generative Approach for Case-Based Reasoning and Prototype Classification," in *NIPS'14: Proceedings of the 22nd International Conference on Neural Information Processing Systems* (2014), 288–296.

14. Kahneman, *Thinking, Fast and Slow*.

15. B. Hayes and J. A. Shah, "Improving Robot Controller Transparency Through Autonomous Policy Explanation," in *2017 ACM/IEEE International Conference on Human-Robot Interaction (HRI)*, Vienna, 2017, 303–312, https://doi.org/10.1145/2909824.3020233.

16. Daniel Szafir, Bilge Mutlu, and Terry Fong, "Communicating Directionality in Flying Robots," in *10th ACM/IEEE International Conference on*

Human-Robot Interaction (HRI), Portland, OR, 2015, 19–26, https://doi.org/10
.1145/2696454.2696475.

CHAPTER 7. ROBOTS TALKING AMONG THEMSELVES

1. *Investigation Report AX001-1-2/02: Boeing B757-200 and Tupolev TU154M, 1 July 2002* (Brunswick, Germany: Bundesstelle für Flugunfalluntersuchung, 2004), https://cfapp.icao.int/fsix/sr/reports/02001351_final_report_01.pdf.

2. James E. Kuchar and Ann C. Drumm, "The Traffic Alert and Collision Avoidance System," *Lincoln Laboratory Journal* 16, no. 2 (2007): 277–296.

3. Paul M. Fitts, ed., *Human Engineering for an Effective Air Navigation and Traffic Control System* (Washington, DC: National Research Council, 1951).

4. Diane Mcruer and Ezra Krendel, "The Man-Machine System Concept," *Proceedings of the IRE* 50, no. 5 (1962): 1117–1123, https://doi.org/10.1109/jrproc.1962.288016.

5. Laurence R. Young, "On Adaptive Manual Control," *Ergonomics* 12, no. 4 (1969): 635–674, https://doi.org/10.1080/00140136908931083; Diane Mcruer and D. H. Weir, "Theory of Manual Vehicular Control," *Ergonomics* 12, no. 4 (1969): 599–633, https://doi.org/10.1080/00140136908931082.

6. Jeff Wise, "Is the Boeing 737 Max Worth Saving?," *New York*, March 29, 2019, https://nymag.com/intelligencer/2019/03/is-the-boeing-737-max-worth-saving.html.

7. "737 Max Updates," Boeing, https://www.boeing.com/commercial/737max/737-max-software-updates.page; "Boeing 737 Max: What Went Wrong?," BBC, April 5, 2019, https://www.bbc.com/news/world-africa-47553174.

8. Lynne Collis and Paul Robins, "Developing Appropriate Automation for Signalling and Train Control on High Speed Railways," in *2001 People in Control: The Second International Conference on Human Interfaces in Control Rooms, Cockpits and Command Centres*, IET Conference Publication No. 482, Manchester, UK, 2001, 255–260.

9. O. Bebek and M. Cenk Cavusoglu, "Intelligent Control Algorithms for Robotic-Assisted Beating Heart Surgery," *IEEE Transactions on Robotics* 23, no. 3 (2007): 468–480, https://doi.org/10.1109/tro.2007.895077.

10. R. Parasuraman, T. B. Sheridan, and C. B. Wickens, "A Model for Types and Levels of Human Interaction with Automation," *IEEE Transactions on Systems, Man, and Cybernetics, Part A: Systems and Humans* 30, no. 3 (2000): 286–297, https://doi.org/10.1109/3468.844354.

11. R. John Hansman, "Complexity in Aircraft Automation—A Precursor for Concerns in Human-Automation Systems," *Phi Kappa Phi National Forum* 81, no. 1 (2001): 30; Michael A. Mollenhauer, Thomas A. Dingus, Cher Carney,

Jonathan M. Hankey, and Steve Jahns, "Anti-Lock Brake Systems: An Assessment of Training on Driver Effectiveness," *Accident Analysis and Prevention* 29, no. 1 (1997): 97–108, https://doi.org/10.1016/s0001-4575(96)00065-6.

12. Mikael Ljung Aust, Lotta Jakobsson, Magdalena Lindman, and Erik Coelingh, "Collision Avoidance Systems—Advancements and Efficiency," *SAE Technical Paper* no. 2015-01-1406 (2015), https://doi.org/10.4271/2015-01-1406.

13. Thomas Sheridan, "Space Teleoperation Through Time Delay: Review and Prognosis," *IEEE Transactions on Robotics and Automation* 9, no. 5 (1993): 592–606, https://doi.org/10.1109/70.258052; Thomas Sheridan, "Teleoperation, Telerobotics and Telepresence: A Progress Report," *Control Engineering Practice* 3, no. 2 (1995): 205–214, https://doi.org/10.1016/0967-0661(94)00078-u; Gary Witus, Shawn Hunt, and Phil Janicki, "Methods for UGV Teleoperation with High Latency Communications," in *Proceedings Volume 8045: SPIE Defense, Security, and Sensing. Unmanned Systems Technology XIII* (March 2011), 80450N, https://doi.org/10.1117/12.886058.

14. Jonathan Bohren, Chris Paxton, Ryan Howarth, Gregory D. Hager, and Louis L. Whitcomb, "Semi-Autonomous Telerobotic Assembly over High-Latency Networks," in *11th ACM/IEEE International Conference on Human-Robot Interaction (HRI)*, Christchurch, 2016, 149–156, https://doi.org/10.1109/hri.2016.7451746; Eric Krotkov, Douglas Hackett, Larry Jackel, Michael Perschbacher, James Pippine, Jesse Strauss, Gill Pratt, and Christopher Orlowski, "The DARPA Robotics Challenge Finals: Results and Perspectives," *Journal of Field Robotics* 34, no. 2 (2016): 229–240, https://doi.org/10.1002/rob.21683.

15. Emily Lakdawalla, *The Design and Engineering of Curiosity: How the Mars Rover Performs Its Job* (New York: Springer International, 2018).

16. Mary L. Cummings and Stephanie Guerlain, "Developing Operator Capacity Estimates for Supervisory Control of Autonomous Vehicles," *Human Factors* 49, no. 1 (2007): 1–15; Gloria L. Calhoun, Michael A. Goodrich, John R. Dougherty, Julie A. Adams, and N. Cooke, "Human-Autonomy Collaboration and Coordination Toward Multi-RPA Missions," in *Remotely Piloted Aircraft Systems: A Human Systems Integration Perspective*, ed. Nancy J. Cooke, Leah J. Rowe, Winston Bennett Jr., and DeForest Q. Joralmon (Chichester, West Sussex, UK: John Wiley and Sons, 2016), 109–136, https://doi.org/10.1002/9781118965900.ch5.

17. Jin-Hee Cho, Yating Wang, Ing-Ray Chen, Kevin S. Chan, and Ananthram Swami, "A Survey on Modeling and Optimizing Multi-Objective Systems," *IEEE Communications Surveys & Tutorials* 19, no. 3 (2017): 1867–1901, https://doi.org/10.1109/comst.2017.2698366; Sameera Ponda, Josh Redding, Han-Lim Choi, Jonathan P. How, Matt Vavrina, and John Vian, "Decentralized Planning for Complex Missions with Dynamic Communication Constraints," in *Proceedings of the 2010 American Control Conference* (Piscataway, NJ: IEEE: 2010), 3998–4003,

https://doi.org/10.1109/acc.2010.5531232; Federico Celi, Li Wang, Lucia Pallot-tino, and Magnus Egerstedt, "Deconfliction of Motion Paths with Traffic Inspired Rules," *IEEE Robotics and Automation Letters* 4, no. 2 (2019): 2227–2234, https://doi.org/10.1109/lra.2019.2899932.

18. Justin Werfel, Kirstin Petersen, and Radhika Nagpal, "Designing Collective Behavior in a Termite-Inspired Robot Construction Team," *Science* 343, no. 6172 (2014): 754–758, https://doi.org/10.1126/science.1245842.

19. Jin-Hee Cho et al., "Survey on Modeling."

CHAPTER 8. THIS CITY IS A CYBORG

1. Sebastian Timar, George Hunter, and Joseph Post, "Assessing the Benefits of NextGen Performance-Based Navigation," *Air Traffic Control Quarterly* 21, no. 3 (2013): 211–232, https://doi.org/10.2514/atcq.21.3.211.

2. *ISO/TS 15066:2016 Robots and Robotic Devices—Collaborative Robots* (Geneva: ISO, 2016), https://www.iso.org/standard/62996.html; Przemyslaw A. Lasota, Gregory F. Rossano, and Julie A. Shah, "Toward Safe Close-Proximity Human-Robot Interaction with Standard Industrial Robots," in *2014 IEEE International Conference on Automation Science and Engineering (CASE)*, Taipei City, 2014, 339–344, https://doi.org/10.1109/coase.2014.6899348; Jeremy A. Marvel and Rick Norcross, "Implementing Speed and Separation Monitoring in Collaborative Robot Workcells," *Robotics and Computer-Integrated Manufacturing* 44 (April 2017): 144–155, https://doi.org/10.1016/j.rcim.2016.08.001; Akansel Cosgun, Emrah Akin Sisbot, and Henrik Iskov Christensen, "Anticipatory Robot Path Planning in Human Environments," in *2016 25th IEEE International Symposium on Robot and Human Interactive Communication (RO-MAN)*, New York, 2016, https://doi.org/10.1109/roman.2016.7745174; Vaibhav V. Unhelkar, Przemyslaw A. Lasota, Quirin Tyroller, Rares-Darius Buhai, Laurie Marceau, Barbara Deml, and Julie A. Shah, "Human-Aware Robotic Assistant for Collaborative Assembly: Integrating Human Motion Prediction with Planning in Time," *IEEE Robotics and Automation Letters (RA-L)* 3, no. 3 (2018): 2394–2401, https://doi.org/10.1109/lra.2018.2812906.

3. "Smarter, Smaller, Safer Robots," *Harvard Business Review*, November 2015, https://hbr.org/2015/11/smarter-smaller-safer-robots.

4. Nick A. Komons, *Bonfires to Beacons: Federal Civil Aviation Policy Under the Air Commerce Act 1926-1938* (Washington, DC: FAA, 1978); Charles F. Horne, "Airways: Today and Tomorrow," *Signals* 7–9 (1953): 11.

5. *Accident Investigation Report, Official Report SA-320, File No. 1-0090*, Civil Aeronautics Board, April 17, 1957, https://lessonslearned.faa.gov/UAL718/CAB_accident_report.pdf.

6. Federal Aviation Act, Public Law 85-726, 85th Cong., 1958, https://www.govinfo.gov/content/pkg/STATUTE-72/pdf/STATUTE-72-Pg731.pdf.

7. *ISO/TS 15066:2016 Robots and Robotic Devices*.

8. *Produktion* 2017 13, March 2017, https://www.produktion.de/abonnement/heftarchiv.html?page=10.

9. *World Robotics Report 2017* (Frankfurt: International Federation of Robotics, 2017).

10. "Smarter, Smaller, Safer Robots."

11. *World Robotics Report 2017*.

12. John Harding, Gregory Powell, Rebecca Yoon, Joshua Fikentscher, Charlene Doyle, Dana Sade, Mike Lukuc, Jim Simons, and Jing Wang, *Vehicle-to-Vehicle Communications: Readiness of V2V Technology for Application*, Report No. DOT HS 812 014 (Washington, DC: National Highway Traffic Safety Administration, August 2014), http://www.nhtsa.gov/staticfiles/rulemaking/pdf/V2V/Readiness-of-V2V-Technology-for-Application-812014.pdf.

13. Sarah Keren, Avigdor Gal, and Erez Karpas, "Goal Recognition Design," in *Twenty-Fourth International Conference on Automated Planning and Scheduling*, Portsmouth, NH, 2014.

CHAPTER 9. IT TAKES A VILLAGE TO RAISE A ROBOT

1. *Highway Accident Report: Collision Between a Car Operating with Automated Vehicle Control Systems and a Tractor-Semitrailer Truck Near Williston, Florida, May 7, 2016*, NTSB/HAR-17/02 (Washington, DC: National Transportation Safety Board, 2017), https://www.ntsb.gov/investigations/AccidentReports/Reports/HAR1702.pdf.

2. Charles Fleming, "Tesla Car Mangled in Fatal Crash Was on Autopilot and Speeding, NTSB Says," *Los Angeles Times*, July 26, 2016, https://www.latimes.com/business/autos/la-fi-hy-autopilot-photo-20160726-snap-story.html.

3. Fred Lambert, "Understanding the Fatal Tesla Accident on Autopilot and the NHTSA Probe," *Electrek*, July 1, 2016.

4. "Autonomous Vehicle Disengagement Reports 2017," State of California Department of Motor Vehicles, https://www.dmv.ca.gov/portal/dmv/detail/vr/autonomous/disengagement_report_2017.

5. Federal Aviation Act, Public Law 85-726, 85th Congress, 1958, https://www.govinfo.gov/content/pkg/STATUTE-72/pdf/STATUTE-72-Pg731.pdf; US Congress, Senate, Subcommittee on Aviation of the Committee on Interstate and Foreign Commerce, Federal Aviation Agency Act: Hearings Before the Subcommittee on Aviation of the Committee on Interstate and Foreign Commerce, 85th Congress, 2nd sess., 1958.

6. Bobbie R. Allen, Flight Safety Foundation International Air Safety Seminar, Madrid, 1966.

7. NASA, *Aviation Safety Reporting System*, July 2019, https://asrs.arc.nasa .gov/docs/ASRS_ProgramBriefing.pdf.

8. "Aviation Safety Reporting System," NASA, https://asrs.arc.nasa.gov.

9. US Congress, House of Representatives, Subcommittee of the Committee on Government Operations, *FAA Aviation Safety Reporting System: Hearing Before a Subcommittee of the Committee on Government Operations*, 96th Cong., 1st sess., 1979.

10. NASA, *Aviation Safety Reporting System*.

11. *ASRS Database Report Set: Wake Turbulence Encounters* (Moffatt Field, NASA Ames Research Center, 2018), https://asrs.arc.nasa.gov/docs/rpsts /waketurb.pdf.

12. "Program Summary," Confidential Close Call Reporting System (C3RS), NASA, https://c3rs.arc.nasa.gov/information/summary.html.

13. George Dyson, *Darwin Among the Machines* (New York: Basic Books, 2012).

14. Jeffrey C. Mogul, "Emergent (Mis)Behavior vs. Complex Software Systems," *ACM SIGOPS Operating Systems Review* 40, no. 4 (January 2006): 293, https://doi.org/10.1145/1218063.1217964.

15. Hans Van Vliet, *Software Engineering: Principles and Practice* (Chichester, West Sussex, UK: John Wiley and Sons, 2008).

16. Jeffrey Mahler, Jacky Liang, Sherdil Niyaz, Michael Laskey, Richard Doan, Xinyu Liu, Juan Aparicio, and Ken Goldberg, "Dex-Net 2.0: Deep Learning to Plan Robust Grasps with Synthetic Point Clouds and Analytic Grasp Metrics," in *Robotics: Science and Systems* 13 (December 2017), https://doi.org/10.15607 /rss.2017.xiii.058.

17. Jonathan Tremblay, Thang To, Artem Molchanov, Stephen Tyree, Jan Kautz, and Stan Birchfield, "Synthetically Trained Neural Networks for Learning Human-Readable Plans from Real-World Demonstrations," in *2018 IEEE International Conference on Robotics and Automation (ICRA)*, Brisbane, 2018, https:// doi.org/10.1109/icra.2018.8460642.

18. Michele Banko and Eric Brill, "Scaling to Very Very Large Corpora for Natural Language Disambiguation," *ACL 01: Proceedings of the 39th Annual Meeting on Association for Computational Linguistics* (New York: ACM, 2001), 26–33, https://doi.org/10.3115/1073012.1073017.

19. Xavier Amatriain, "In Machine Learning, What Is Better: More Data or Better Algorithms?," *KDnuggets*, June 2015, https://www.kdnuggets.com/2015/06 /machine-learning-more-data-better-algorithms.html.

20. Joy Buolamwini and Timnit Gebru, "Gender Shades: Intersectional Accuracy Disparities in Commercial Gender Classification," in *2018 Conference on Fairness, Accountability and Transparency* (New York: ACM, 2018), 77–91.

21. Timnit Gebru, Jamie Morgenstern, Briana Vecchione, Jennifer Wortman Vaughan, Hanna Wallach, Hal Daumé III, and Kate Crawford, "Datasheets for Datasets," *arXiv:1803.09010* (2018).

22. Solace Shen, Hamish Tennent, Houston Claure, and Malte Jung, "My Telepresence, My Culture?," in *Proceedings of the 2018 CHI Conference on Human Factors in Computing Systems*, Montreal, 2018, 1–11, https://doi.org/10.1145/3173574.3173625.

23. "ImageNet," Stanford Vision Lab, Stanford University, Princeton University, http://www.image-net.org.

24. "NuScenes Dataset," Aptiv, https://www.nuscenes.org.

25. Jie Hu, Li Shen, and Gang Sun, "Squeeze-and-Excitation Networks," in *2018 IEEE/CVF Conference on Computer Vision and Pattern Recognition* (Piscataway, NJ: IEEE, 2018), 7132–7141, https://doi.org/10.1109/cvpr.2018.00745.

26. NASA, *Aviation Safety Reporting System*; "Status Report and FY05 Funding Impacts," Meeting Minutes for a Briefing to the Aviation Safety Reporting System Subcommittee, NASA, November 3, 2004, https://www.hq.nasa.gov/office/aero/advisors/asrss/11_03_04/funding.htm.

Index

LAURA MAJOR is the CTO of Motional, where she leads the development of autonomous vehicles. Previously, she led the development of autonomous aerial vehicles at CyPhy Works and a division at Draper Laboratory. Major has been recognized as a national Society of Women Engineers Emerging Leader. She lives in Cambridge, Massachusetts. *Photo courtesy of the author.*

JULIE SHAH is an associate professor of aeronautics and astronautics and associate dean for social and ethical responsibilities of computing at the Massachusetts Institute of Technology, where she also directs the Interactive Robotics Group. Shah has been recognized by the National Science Foundation with a Faculty Early Career Development (CAREER) award and by *MIT Technology Review* on its Innovators Under 35 list. Her work on industrial human-robot collaboration was also in *MIT Technology Review*'s 2013 list of 10 Breakthrough Technologies. She lives in Cambridge, Massachusetts. *Photo courtesy of Dennis Kwan.*